Lectures on Astrophysics

Lectures on Astrophysics provides an account of classic and contemporary aspects of astrophysics, with an emphasis on analytic calculations and physical understanding. It introduces fundamental topics in astrophysics, including the properties of single and binary stars, the phenomena associated with interstellar matter, and the structure of galaxies. Nobel Laureate Steven Weinberg combines exceptional physical insight with his gift for clear exposition to cover exciting recent developments and new results. Emphasizing theoretical results, and explaining their derivation and application, this book provides an invaluable resource for physics and astronomy students and researchers.

STEVEN WEINBERG is a Professor of Physics and Astronomy at the University of Texas at Austin. His research has covered a broad range of topics in quantum field theory, elementary particle physics, and cosmology, and he has been honored with numerous awards, including the Nobel Prize in Physics, the National Medal of Science, and the Heinemann Prize in Mathematical Physics. The American Philosophical Society awarded him the Benjamin Franklin medal, with a citation that said he is "considered by many to be the preeminent theoretical physicist alive in the world today." He is a member of several academies in the USA and abroad, including the US National Academy of Sciences and Britain's Royal Society. He has written several highly regarded books, including *Gravitation and Cosmology* (Wiley, 1972), a three-volume work *The Quantum Theory of Fields* (CUP, 2005), *Cosmology* (OUP, 2008), and *Lectures on Quantum Mechanics* (CUP, 2nd edn., 2015).

Lectures on Astrophysics

Steven Weinberg
University of Texas, Austin

CAMBRIDGE
UNIVERSITY PRESS

University Printing House, Cambridge CB2 8BS, United Kingdom

One Liberty Plaza, 20th Floor, New York, NY 10006, USA

477 Williamstown Road, Port Melbourne, VIC 3207, Australia

314–321, 3rd Floor, Plot 3, Splendor Forum, Jasola District Centre, New Delhi – 110025, India

79 Anson Road, #06-04/06, Singapore 079906

Cambridge University Press is part of the University of Cambridge.

It furthers the University's mission by disseminating knowledge in the pursuit of education, learning, and research at the highest international levels of excellence.

www.cambridge.org
Information on this title: www.cambridge.org/9781108415071
DOI: 10.1017/9781108227445

© Cambridge University Press 2020

This publication is in copyright. Subject to statutory exception and to the provisions of relevant collective licensing agreements, no reproduction of any part may take place without the written permission of Cambridge University Press.

First published 2020

Printed in the United Kingdom by TJ International Ltd, Padstow, Cornwall

A catalogue record for this publication is available from the British Library.

Library of Congress Cataloging-in-Publication Data
Names: Weinberg, Steven, 1933– author.
Title: Lectures on astrophysics / Steven Weinberg (University of Texas, Austin).
Description: Cambridge ; New York, NY : Cambridge University Press, 2019. | Includes bibliographical references.
Identifiers: LCCN 2019021306 | ISBN 9781108415071 (hardback)
Subjects: LCSH: Astrophysics.
Classification: LCC QB461 .W375 2019 | DDC 523.01–dc23
LC record available at https://lccn.loc.gov/2019021306

ISBN 978-1-108-41507-1 Hardback

Cambridge University Press has no responsibility for the persistence or accuracy of URLs for external or third-party internet websites referred to in this publication and does not guarantee that any content on such websites is, or will remain, accurate or appropriate.

For Louise, Elizabeth, and Gabrielle

Contents

PREFACE *page* xiii

1   STARS   1

1.1   Hydrostatic Equilibrium   2
Equilibrium equation □ Central pressure □ Gravitational binding energy
□ Virial theorem □ Stability □ Initial contraction □ Kelvin time scale

1.2   Radiative Energy Transport   7
Differential energy density □ Transport term □ Absorption term
□ Scattering term □ Emission term □ Equilibrium □ Flux divergence
□ Momentum tensor divergence □ Opacity □ Rosseland mean
□ Radiative transport equations

1.3   Radiative Models   14
Differential equations □ Conditions at center □ Conditions at
nominal surface □ True surface □ Vogt–Russell theorem
□ Effective temperature □ Color temperature □ Hertzsprung–Russell relation
□ Eddington bound

1.4   Opacity   21
Contributions to opacity □ Stimulated emission □ Thomson scattering
□ Free–free absorption □ Kramers opacity □ Bound–free absorption
□ Bound–bound absorption □ *Appendix*: Calculation of free–free absorption

1.5   Nuclear Energy Generation   30
Proton–proton chain □ CNO cycle □ Suppression factors
□ Coulomb barrier □ Application to proton–proton chain □ Solar neutrinos

□ Application to CNO cycle □ Crossover □ Beyond hydrogen burning
□ Carbon synthesis □ *Appendix*: Calculation of suppression by
Coulomb barriers

1.6 Relations among Observables: The Main Sequence 42

Temperature and density dependence of energy generation and opacity
□ Dimensional analysis □ Gas pressure dominance: radius–mass relation,
luminosity–mass relation, central temperature versus effective surface temperature,
Hertzsprung–Russell slope □ Hydrogen burning time □ Radiation pressure
dominance: radius–mass relation, luminosity–mass relation,
Hertzsprung–Russell slope

1.7 Convection 50

Stability against convection □ Eddington discriminant
□ Mixing length theory □ Efficient convection □ Isentropic stars
□ The Sun □ Variational principle

1.8 Polytropes 59

Examples of polytropic stars □ The Lane–Emden differential equation
□ Exact solutions □ Numerical solutions

1.9 Instability 63

Onset of instability: general theorem, with exceptions □ Stars close to $\Gamma = 4/3$
□ Expansion in $1/c^2$ □ *Appendix*: Derivation of relativistic energy correction

1.10 White Dwarfs and Neutron Stars 70

Equation of state for cold electrons □ High-mass and low-mass white dwarfs
□ Neutronization □ Relativistic instability □ Equation of state for
cold neutrons □ Low-mass neutron stars □ Landau–Oppenheimer–Volkoff limit
□ Neutron star spin □ Pulsars

1.11 Supermassive Stars 77

Gas/radiation pressure ratio □ Equation of state □ Mass
□ Stability □ Evolution

Bibliography for Chapter 1 80

2   BINARIES 82

2.1 Orbits 82

General orbits □ Spectroscopic binaries □ Energy and angular momentum
□ Relativistic corrections □ *Appendix*: Calculation of time dilation in binary stars

2.2 Close Binaries 86
Roche limit □ Sirius A and B □ Equipotential surfaces □ Roche lobes □ Mass transfer □ Type 1a supernovae □ Roche lobe volumes

2.3 Gravitational Wave Emission: Binary Pulsars 92
The Hulse–Taylor pulsar □ Quadrupole approximation for emitted power □ Decrease in period □ Decrease in eccentricity □ Time to coalescence □ Gamma ray bursts and kilonovae □ Total radiated energy □ More binary pulsars □ *Appendix*: Review of gravitational radiation

2.4 Gravitational Wave Detection: Coalescing Binaries 104
Weber bars □ Interferometers □ Sources □ Black holes versus neutron stars □ Chirps □ Description of LIGO □ Transformation to transverse-traceless gauge □ Response of LIGO to gravitational waves □ Shot noise and seismic noise □ Sensitivity □ 2015 detection of gravitational wave □ Diagnosis of source: chirp mass, relativistic corrections □ Estimate of distance □ More coalescing black-hole binaries □ A coalescing neutron star binary □ Blind spots

Bibliography for Chapter 2 116

3 THE INTERSTELLAR MEDIUM 117

3.1 Spectral Lines 118
General transport equation □ Optical depth □ Solution for homogeneous emission/absorption ratio □ Doppler broadening □ Einstein A and B coefficients □ Emission lines from clouds in thermal equilibrium □ Emission lines from non-equilibrium regions □ Absorption lines □ 21 cm lines □ CN absorption lines

3.2 HII Regions 126
Strömgren spheres □ Differential equation for ionization □ Interior of the sphere □ Surface of the sphere □ Recombination lines □ Heating

3.3 Cooling 131
Cooling function □ Prompt radiation case □ Excitation by electrons □ Hydrogen atoms □ Russell–Saunders classification of atom and ion states □ CII □ OIII □ OII □ Cooling in HII regions □ Delayed radiation case □ H_2 and CO molecules □ Bremsstrahlung cooling

3.4 Star Formation 143
Virial estimates ☐ Jeans radius and mass ☐ Molecular clouds
☐ Dispersion relation for gravitational perturbations
☐ Transition to instability ☐ Collapse time

3.5 Accretion Disks 148
Exceeding the Eddington limit ☐ Role of viscosity ☐ Differential equations for surface density ☐ Mass and angular-momentum flow ☐ Steady disks ☐ Viscous heating ☐ Spectral distribution ☐ Thickness of disk ☐ Decaying disks ☐ Bessel function solution for constant viscosity ☐ Expansion of disk ☐ Accretion disks in binaries ☐ Cataclysmic variables

3.6 Accretion Spheres 160
Bondi accretion ☐ Conservation laws ☐ The wind equation ☐ Transonic solutions ☐ Mass accretion rate ☐ M31*

3.7 Soft Bremsstrahlung 164
Emissivity and Gaunt factor ☐ Born approximation ☐ A misleading formula ☐ Low-energy theorem ☐ Debye screening

Bibliography for Chapter 3 171

4 GALAXIES 172

4.1 Collisionless Dynamics 172
Collisionless Boltzmann equation ☐ Surface density from velocity dispersion ☐ Moment equations ☐ Solutions to Boltzmann equation ☐ Eddington theorem

4.2 Polytropes and Isothermals 178
Polytrope solutions of Boltzmann equation ☐ Isothermal solutions of Boltzmann equation ☐ Galaxy clusters ☐ Dark matter ☐ Missing baryons? ☐ NFW distribution

4.3 Galactic Disks 185
Rotation curves ☐ Bulge dominance ☐ Disk dominance ☐ Halo dominance ☐ *Appendix A*: Gravitational potential of a disk ☐ *Appendix B*: Minimum energy configuration for fixed angular momentum

4.4 Spiral Arms 191
Trailing and leading spirals ☐ Differential equations for surface density ☐ Lin–Shu density waves ☐ Winding from differential rotation ☐ Pitch angle and winding problem ☐ Epicyclic frequency ☐ Pattern frequency ☐ Crowding ☐ Lindblad resonances

4.5 Quasars 199

Quasi-stellar objects and sources □ Accretion on black holes □ Heating of accretion disks □ *Appendix*: Orbits of minimum radius about black holes

Bibliography for Chapter 4 203

ASSORTED PROBLEMS 204

AUTHOR INDEX 207

SUBJECT INDEX 211

Preface

This book grew out of the lecture notes for a course on astrophysics that I gave in Austin in the spring term of 2016 and again in the fall term of 2017. In contrast to other courses I have given over the years on general relativity and cosmology, in this course I wanted to provide an introduction to the more traditional "nuts and bolts" aspects of astrophysics: the properties of single and binary stars, the phenomena associated with interstellar matter, and the structure of galaxies.

This is not a comprehensive account of astrophysics and its applications to astronomy. That would not be possible in a single volume. Indeed, many of the individual topics treated in single sections of this book have been the subjects of massive monographs, on which I have heavily relied, especially for numerical results and summaries of astronomical data. (Some of these books are listed among others in bibliographies for each chapter.) Instead, I here offer a short course, a collection of astrophysical calculations that can be done simply and analytically, without recourse to computers, and yet are relevant to the real world.

In many of the treatises and review articles on astrophysics that I have consulted, where numerical computations have not entirely replaced analysis, formulas are given without presenting derivations or even references. (Sometimes they are wrong.) This book is intended for those who care about the rationale of astrophysical formulas as well as about their applications. So where I can I give derivations of all the formulas I use, or if that would take me too far from my subject I give a reference where the derivation can be found.

There are several results presented here that I have not seen elsewhere. Some of them may be new, including a general formula for the matrix element for bremsstrahlung and inverse bremsstrahlung, a formula for the volume of the Roche lobe when one star in a binary is much more massive than the other, and an estimate of the ratio of the central temperature and effective surface temperature of a main sequence star in terms of the center's optical depth.

This is not intended as a treatise on general relativity or cosmology. Having written such books in the past, I had no wish now to write another one. But it would have been impossible to leave out general relativity altogether. For one thing, that would have precluded any serious discussion of the instability of white dwarfs, neutron stars, and supermassive stars. It would have ruled out an estimate of energy production in accretion disks around massive black holes. Above all, I could hardly leave out any discussion of the exciting new field of gravitational wave astronomy, which begin in late 2015 with the discovery by the Laser Interferometric Gravitational Wave Observatory of gravitational waves from distant coalescing black holes.

So I have compromised. In those sections (not many) where general relativity has to be used, brief appendices provide a compact account of relevant aspects of the subject. Derivations are generally abbreviated or skipped where they were given in my 1972 treatise, *Gravitation and Cosmology*, and where topics were not covered there, I take this opportunity to bring that book up to date. Likewise, where I need to use quantum mechanics, for instance to calculate Coulomb barriers in the section on nuclear energy generation or inverse bremsstrahlung in the section on opacity, the calculations are presented in appendices to those sections. I hope that with these appendices, the material presented in this book should be accessible to anyone with a good undergraduate background in classical physics and its mathematical methods.

Of course, as progress is made in observational astronomy new calculations will become relevant, but I trust that the calculations presented here will provide physicists and astrophysicists with a kit of analytic tools of permanent value.

In preparing these lectures, I have greatly benefited from conversations with many physicists, astrophysicists, and astronomers. Special thanks are due to colleagues at the University of Texas: Michael Boylan-Kolchin, Richard Matzner, Paul Shapiro, J. Craig Wheeler, and Aaron Zimmerman. I am very grateful to Anson d'Aloisio and to Aaron Smith for reading through the book's first draft, and making numerous valuable suggestions. Any errors that remain are all my fault. As with my earlier book on quantum mechanics, I owe many thanks to Simon Capelin of Cambridge University Press for his help in bringing out this volume.

Many years ago when I was bed-ridden in Berkeley with a bad back, my wife gave me a present, a copy of Chandrasekhar's 1939 classic *An Introduction to the Study of Stellar Structure* that she had found in a bookshop on Telegraph Avenue. Reading the book saved me from wasting much of my time in bed, and gave me a permanent sense of excitement that physics and mathematics could deal effectively with something as mysterious as stars. I don't wish bad backs on today's young physicists, but I hope that they will have some occasion to spend time going through these calculations, and will feel some of the excitement with astrophysics that I first felt long ago.

STEVEN WEINBERG

1
Stars

In antiquity stars were generally supposed to be bright spots fixed on a sphere that revolves once a day about the Earth. Modern astrophysics began in the early nineteenth century, with the discovery by Joseph von Fraunhofer (1787–1826) of dark lines in the spectra of the Sun and some bright stars, which showed that they all have similar composition, and with the measurement by Friedrich Bessel (1766–1828) and William Wollaston (1784–1826) of the distances of stars like Y Cygni and α Centauri, which showed that their absolute luminosity is not very different from that of the Sun. By the end of the nineteenth century hydrodynamics and thermodynamics had been applied to the structure of the Sun and stars. Only the source of their energy was still mysterious, not to be understood in detail until the development of nuclear physics in the 1930s.

It would be most logical to begin this chapter with an introduction to the physics required to understand modern stellar theory, including calculations of nuclear energy production and opacity, and only then go on to the stars themselves. Logical, but perhaps a bit boring. It is not always possible to maintain one's interest in the details of nuclear and atomic physics without knowing how these results are to be used. So in this chapter we start with the stars.

First in Section 1.1 we derive the equations of hydrostatic equilibrium for stars. This leads to the virial theorem, which illuminates the stars' early history. Then in Sections 1.2 and 1.3 we adopt a simple model in which energy is transported in the star solely by radiation, leaving convection for later sections. In this model we can see how the structure of the star is uniquely determined by the formulas that give pressure, opacity, and nuclear energy production in terms of density and temperature, with just one free stellar parameter, that can be taken to be the star's total mass. With this as motivation, in Sections 1.4 and 1.5 we describe the physics underlying the formulas for opacity and nuclear energy generation. It turns out to be a fair approximation to take the opacity and energy generation as well as the pressure as proportional to products of powers of density and temperature. This approximation is used in Section 1.6 to give formulas for stellar properties, including luminosity, radius, central temperature, etc., in terms of the star's mass. We come to convection in Section 1.7, and show that the presence of convective zones does not greatly change the results of Sections 1.3 and 1.6.

We then turn to stars of a more exotic breed. In Section 1.8 we consider the general class of stars in which the pressure is simply proportional to some power of the density. Where this power is close to 4/3, the star is close to instability. The detailed conditions for stellar instability are worked out in Section 1.9. Then we consider white dwarf and neutron stars in Section 1.10 and supermassive stars in Section 1.11, using the results of Section 1.8 to describe their structure and of Section 1.9 to find where they become unstable.

This chapter deals only with isolated single stars. Binary stars and their emission of gravitational radiation will be considered in the following chapter.

1.1 Hydrostatic Equilibrium

Suppose a star is in equilibrium and is spherically symmetric, so that the mass density ρ and pressure p are functions only of the distance r from the center. Consider a thin spherical shell of radius r and thickness dr. Its mass is $4\pi r^2 \rho(r)\, dr$, so it feels a gravitational force

$$F_{\text{gravitational}} = -G \frac{4\pi r^2 \rho(r)\, dr\, \mathcal{M}(r)}{r^2} = -4\pi G \rho(r) \mathcal{M}(r)\, dr, \qquad (1.1.1)$$

where $\mathcal{M}(r)$ is the total mass interior to the radius r:

$$\mathcal{M}(r) = \int_0^r 4\pi r'^2 \rho(r')\, dr'. \qquad (1.1.2)$$

The minus sign in Eq. (1.1.1) indicates that this force points inward. The shell also feels an outward buoyant force, equal to the pressure force on the inner surface of the shell minus the pressure force on its outer surface:

$$F_{\text{buoyant}} = 4\pi r^2 \big[p(r) - p(r+dr)\big] = -4\pi r^2\, p'(r)\, dr. \qquad (1.1.3)$$

In equilibrium the sum of these forces vanishes, so

$$\frac{dp(r)}{dr} = -\frac{G\mathcal{M}(r)\rho(r)}{r^2}. \qquad (1.1.4)$$

This is the fundamental equation of hydrostatic equilibrium for stars. For some purposes it is convenient to rewrite Eq. (1.1.2) also as a differential equation

$$\frac{d\mathcal{M}(r)}{dr} = 4\pi r^2 \rho(r), \qquad (1.1.5)$$

with initial condition $\mathcal{M}(0) = 0$.

Equations (1.1.4) and (1.1.5) lead to a useful inequality for the pressure.[1] We note that

$$\frac{d}{dr}\left[p(r) + \frac{G\mathcal{M}^2(r)}{8\pi r^4}\right] = -\frac{G\mathcal{M}(r)\rho(r)}{r^2} - \frac{G\mathcal{M}^2(r)}{2\pi r^5} + \frac{G\mathcal{M}(r)\mathcal{M}'(r)}{4\pi r^4}.$$

The first and third terms cancel, leaving the negative second term, so

$$\frac{d}{dr}\left[p(r) + \frac{G\mathcal{M}^2(r)}{8\pi r^4}\right] \leq 0. \tag{1.1.6}$$

In particular, assuming that the density is finite at $r = 0$, we have $\mathcal{M}(r) \propto r^3$ for $r \to 0$, so $\mathcal{M}^2(r)/r^4 \to 0$ for $r \to 0$. Assuming also that the pressure vanishes at some nominal stellar radius R, and taking $\mathcal{M}(R) = M$, the quantity in square brackets in (1.1.6) is $p(0)$ at $r = 0$ and $GM^2/8\pi R^4$ at $r = R$, so (1.1.6) yields a useful inequality for the central pressure:

$$p(0) \geq \frac{GM^2}{8\pi R^4} = 4.44 \times 10^{14}(M/M_\odot)^2(R/R_\odot)^{-4} \,\text{dyne/cm}^2. \tag{1.1.7}$$

(The subscript $\odot$ denotes values for the Sun. For comparison, recall that one standard atmosphere equals 1.013×10^6 dyne/cm^2.) Using methods described in this chapter, it has been calculated that the pressure at the center of the Sun is $p_\odot(0) \simeq 2 \times 10^{17}$ dyne/cm^2, in accord with the inequality (1.1.7).

Equation (1.1.4) can be used to derive a simple formula for the total gravitational potential energy Ω of the star, related to the virial theorem of celestial mechanics. We define $-\Omega$ as the energy required to remove the mass of the star to infinity, peeling it shell by shell from the outside in. Once all the mass exterior to a radius r has been removed, the energy required to remove the shell at r of thickness dr is the integral over the distance r' between the shell and the star's center of the gravitational force $G\mathcal{M}(r)/r'^2 \times 4\pi r^2 \rho(r)\,dr$ exerted by a mass $\mathcal{M}(r)$ on the shell's mass:

$$G\mathcal{M}(r) \times 4\pi r^2 \rho(r)\,dr \times \int_r^\infty \frac{dr'}{r'^2} = 4\pi Gr\mathcal{M}(r)\rho(r)\,dr,$$

so the total gravitational binding energy is

$$-\Omega = 4\pi G \int_0^R r\mathcal{M}(r)\rho(r)\,dr, \tag{1.1.8}$$

where R is the radius of the nominal stellar surface, where $p(R) = 0$. Using Eq. (1.1.4) for $-G\mathcal{M}\rho$, we have

---

[1] S. Chandrasekhar, *An Introduction to the Study of Stellar Structure* (University of Chicago Press, Chicago, IL, 1939), Chapter III. This chapter also gives other general theorems derived from Eqs. (1.1.4) and (1.1.5).

$$\Omega = 4\pi \int_0^R \frac{dp(r)}{dr} r^3 \, dr = -3 \int_0^R p(r) 4\pi r^2 \, dr, \qquad (1.1.9)$$

in which we have integrated by parts, using the vanishing of $r^3 p(r)$ at both endpoints of the integral.

Incidentally, the definition of Ω can also be written in terms of the familiar gravitational potential

$$\phi(r) = -G \int_r^\infty \mathcal{M}(r') \, dr'/r'^2. \qquad (1.1.10)$$

(This formula satisfies the defining condition that $-\phi'(r)$ should equal the Newtonian force per mass $-G\mathcal{M}(r)/r^2$. An arbitrary additive constant has been chosen so that $\phi(r) \to 0$ for $r \to \infty$.) Integrating by parts, we have

$$\int_0^\infty \phi(r) \mathcal{M}'(r) \, dr = -\int_0^\infty \phi'(r) \mathcal{M}(r) \, dr = -G \int_0^\infty \mathcal{M}^2(r) dr/r^2.$$

With $-1/r^2 = d/dr(1/r)$ and integrating by parts again, we see that the final expression is 2Ω, so

$$\Omega = \frac{1}{2} \int_0^\infty \phi(r) \mathcal{M}'(r) \, dr. \qquad (1.1.11)$$

The integral here is the sum of the gravitational energies of each bit of stellar matter, due to the gravitational field of each bit of matter, so in the integral each bit of stellar matter is counted twice, a double counting corrected by the factor $1/2$.

The total energy of the star is the sum of Ω and the star's thermal energy, given by

$$\Upsilon \equiv \int_0^R \mathcal{E}(r) 4\pi r^2 \, dr, \qquad (1.1.12)$$

where $\mathcal{E}(r)$ is the density of internal thermal energy, not including rest mass energies or gravitational energy. The total non-relativistic energy (not including rest masses) of the star is then

$$E = \Upsilon + \Omega = \int_0^R \left[\mathcal{E}(r) - 3p(r)\right] 4\pi r^2 \, dr. \qquad (1.1.13)$$

We see that the star has negative energy and is therefore stable against dispersal of its matter to infinity if $\mathcal{E}(r) < 3p(r)$.

It is frequently the case that the density $\mathcal{E}$ of internal energy is proportional to the pressure, a relation conventionally written as

$$\mathcal{E} = p/(\Gamma - 1). \qquad (1.1.14)$$

(Such stars are called *polytropes*, and are discussed in detail in Section 1.8.) For instance, for an ideal gas of monatomic particles with number density n we have $p = nk_BT$ and $\mathcal{E} = 3nk_BT/2$ (where k_B is Boltzmann's constant), so here $\Gamma = 5/3$. For radiation $p = \mathcal{E}/3$, so $\Gamma = 4/3$. In such cases, the thermal and gravitational energies of the star are given in terms of its total non-relativistic energy by Eqs. (1.1.9), (1.1.12), and (1.1.14) as

$$\Upsilon = -\frac{E}{3\Gamma - 4}, \quad \Omega = \frac{(\Gamma - 1)E}{\Gamma - 4/3}. \tag{1.1.15}$$

The star will explode if E is positive, so stability requires that $E < 0$, and since Eq. (1.1.9) gives $\Omega < 0$, this means that $\Gamma > 4/3$. Stars whose pressure is dominated by highly relativistic particles (such as very massive ordinary stars and white dwarfs and neutron stars with masses near their upper limit) have Γ only slightly above $4/3$ and are therefore trembling on the brink of instability.

Equation (1.1.15) plays a crucial role in governing the early history of stars. A cloud of cold diffuse gas will have little internal or gravitational energy, so its total energy E will be small. Unless the cloud is at absolute zero temperature it will radiate some light, chiefly at infrared wavelengths. If its total energy becomes negative, the cloud will no longer be able to disperse. According to Eq. (1.1.15), as the cloud loses energy then, as long as $\Gamma > 4/3$, Ω will decrease, becoming increasingly negative, but *the internal energy Υ will increase*. The star behaves as if it has negative specific heat; the more it loses energy, the hotter it gets. With increasing temperature the star radiates energy more rapidly, and the process accelerates. Eventually the central temperature of the star becomes so high that nuclei can penetrate the Coulomb repulsion that separates them (discussed in Section 1.5); nuclear energy generation begins and increases until it balances the energy lost by radiation; and the star becomes stable, at least until the nuclear fuel at the star's center is exhausted. Paradoxically, the onset of nuclear reactions *stops* the heating of the star.

As a protostar radiates energy and heats up, it also contracts. We can define a mass-weighted mean radius $\bar{r}$, by

$$M^2 \bar{r}^{-1} \equiv \int_0^R r \mathcal{M}(r) \rho(r) \, dr.$$

Then Eq. (1.1.8) may be written $\Omega = -4\pi G M^2/\bar{r}$. As $-\Omega$ increases, $\bar{r}$ must decrease.

Before the discovery of radioactivity, with its implications for the source of heat of stars, William Thomson (1824–1907, a.k.a. Lord Kelvin), estimated the length of time that the Sun could have been shining with its present luminosity, deriving its heat solely from gravitational contraction.[2] As we have seen, the

---

[2] W. Thomson, *Phil. Mag.* **23**, 158 (1862); reprinted in *Mathematical and Physical Papers by Sir William Thomson, Baron Kelvin*, ed. J. Larmor (Cambridge University Press, Cambridge, 1911).

energy E of a star is related to its gravitational energy Ω by Eq. (1.1.15), which for $\Gamma = 5/3$ gives

$$E = \Omega/2. \qquad (1.1.16)$$

We can get a fair estimate of Ω by taking $\rho(r)$ constant in Eq. (1.1.6), so that $\rho(r) = 3M/4\pi R^3$ and $\mathcal{M}(r) = Mr^3/R^3$, in which case

$$E \simeq -\frac{1}{2} \times 4\pi G \times \frac{MR^2}{5} \times \frac{3M}{4\pi R^3} = -\frac{3GM^2}{10R}. \qquad (1.1.17)$$

This is minus the energy the star has lost in contracting from a cloud with negligible gravitational and thermal energy, if no internal energy sources have contributed to its heat since the contraction began. For the Sun, $M_\odot = 1.9891 \times 10^{33}$ g and $R = 6.960 \times 10^{10}$ cm, so $E \simeq -1.1 \times 10^{48}$ ergs. The Sun's present luminosity is $L_\odot = 3.9 \times 10^{33}$ erg/sec, so in the absence of internal energy sources it could only have been shining at that rate for roughly $|E|/L_\odot \simeq 10^7$ years.[3] Kelvin's 1862 conclusion was not very different: "It seems therefore most probable that the sun has not illuminated the earth for 100,000,000 years." Already in the nineteenth century it was known that this was too short a time for the evolution of life and of features of the Earth's surface, but the path to a resolution of the problem first appeared with the discovery of nuclear energy in 1897.

(By the way, this calculation is sometimes done setting the energy radiated during the Sun's previous life equal to $|\Omega|$ rather than to $|E|$. This ignores the fraction of the energy of gravitational contraction that goes into heating the Sun. As we have seen, that fraction is given by the virial theorem as $1/2$ for $\Gamma = 5/3$, so the Sun's age calculated here is reduced by a factor $1/2$. This serves to emphasize the peculiar aspect of gravitation mentioned above, that as a young star condenses under the influence of gravitation without the production of nuclear energy, it heats up, so that the temperature of a gravitationally condensing body increases as it loses energy.)

In some cases, such as zero-temperature white dwarf stars, the pressure p is a known function of the mass density ρ, which otherwise depends only on chemical composition and universal constants such as $\hbar$, c, and m_e. (This is discussed in Section 1.10.) In such cases, Eqs. (1.1.4) and (1.1.5) yield a definite stellar model.

More generally $p(r)$ depends on the temperature at r as well as on $\rho(r)$, and so Eqs. (1.1.4) and (1.1.5) do not in themselves lead to any definite result for the structure of a star. For this, we also need to understand how energy is transported in the star. There are two chief mechanisms for energy transport, radiation and convection, to be studied in the following sections.

---

[3] See e.g. C. J. Hansen, S. D. Kawaler, and V. Trimble, *Stellar Interiors*, 2nd edn. (Springer, New York, 2004).

1.2 Radiative Energy Transport

The equations of hydrostatic equilibrium involve the pressure, which depends on the temperature, so in order to use them we need equations of energy transport, that dictate how the temperature varies through the star. There are two chief mechanisms of energy transport: radiation and convection. (Because mean free paths are small in stars, conduction is much less important.) In this section we shall work out the coupled differential equations, Eqs. (1.2.28) and (1.2.30), that govern the r-dependent temperature and luminosity for a star in which energy transport is dominated by radiation. Convection will be considered in Section 1.7.

Let $\ell(\hat{n}, \mathbf{x}, \nu, t) \, d^2\hat{n} \, d\nu$ be the energy per volume at position $\mathbf{x}$ and time t of photons with directions within a solid angle $d^2\hat{n}$ around the unit vector $\hat{n}$ and frequencies between ν and $\nu + d\nu$. Our first task is to calculate various contributions to the rate of change of $\ell(\hat{n}, \mathbf{x}, \nu, t)$. Later we shall assume that the total rate of change of $\ell(\hat{n}, \mathbf{x}, \nu, t)$ vanishes, and use that requirement as the condition of equilibrium when energy transport is dominated by radiation.

There are four contributions to this rate of change.

Transport

If nothing is happening to the radiation, then at time $t+dt$ the energy of photons per volume, per solid angle, and per frequency interval traveling in direction $\hat{n}$ with frequency ν at position $\mathbf{x}$ will be what it was at time t and position $\mathbf{x} - c\hat{n} \, dt$:

$$\ell(\hat{n}, \mathbf{x}, \nu, t + dt) = \ell(\hat{n}, \mathbf{x} - c\hat{n} \, dt, \nu, t).$$

Thus the rate of change of ℓ solely due to the transport of radiation is

$$\left(\frac{\partial}{\partial t} \ell(\hat{n}, \mathbf{x}, \nu, t) \right)_{\text{transport}} = -c\hat{n} \cdot \nabla \ell(\hat{n}, \mathbf{x}, \nu, t). \quad (1.2.1)$$

Absorption

It is important to distinguish here between absorption and scattering. We will understand absorption to be any process in which an incident photon disappears without producing a photon whose direction is correlated with that of the incident photon. For instance, in a so-called *bound–free* transition, a photon gives its energy to raising the energy of a bound electron so that it becomes a free particle. In a *free–free* transition the incident photon is absorbed by a free electron in the Coulomb field of an ion (which allows such a transition to conserve energy and momentum). In either case the final free electron merges with the surrounding medium, increasing its temperature. The medium may then give up this energy by emitting photons, but the directions of these photons will be uncorrelated with the initial photon's direction, so these transitions count as absorption. In a *bound–bound* transition the energy of the initial photon goes to raise the atom to a higher energy state. Typically the atom then undergoes collisions, which either drain the excitation energy or change the excited state so that even if it decays radiatively the final photon direction is uncorrelated with the direction of the initial photon. In either case, these bound–bound transitions also count as absorption.

Suppose that the net fraction of radiation of frequency ν absorbed at position $\mathbf{x}$ and time t in a time interval dt is $c\kappa_{\text{abs}}(\mathbf{x}, \nu, t)\rho(\mathbf{x}, t)\,dt$, where ρ is the mass density and κ_{abs} is a coefficient characterizing the medium, called the *absorption opacity*. (As discussed in Section 1.4, stimulated emission counts here as negative absorption.) A factor of the speed of light is inserted here to give $1/\kappa_{\text{abs}}\rho$ the dimensions of length; it is the average distance that a typical photon travels before being absorbed in a homogeneous medium. Then the rate of change of ℓ due to absorption is

$$\left(\frac{\partial}{\partial t}\ell(\hat{n}, \mathbf{x}, \nu, t)\right)_{\text{absorption}} = -c\kappa_{\text{abs}}(\mathbf{x}, \nu, t)\rho(\mathbf{x}, t)\ell(\hat{n}, \mathbf{x}, \nu, t). \qquad (1.2.2)$$

For a two-body absorption process like a bound–free or bound–bound transition $\kappa_{\text{abs}}\rho$ is the absorption cross section times the number density of absorbers, and hence κ_{abs} is the absorption cross section divided by the mean absorber mass. (As we will see in Section 1.4, free–free transitions are more complicated.)

Scattering

These are processes in which the disappearance of an initial photon yields a final photon, whose direction generally differs from the initial direction, but is correlated with it. The leading example is Thomson scattering, the elastic scattering of photons with energies well below $m_e c^2$ on non-relativistic electrons. A bound–bound transition could also be regarded as a scattering, if the excited atom were to decay radiatively before the atom undergoes collisions that wipe out any correlation of the final and initial photons.

The fraction of radiation energy of frequency ν traveling in a direction $\hat{n}$ that in a time interval dt at time t is scattered at position $\mathbf{x}$ into a solid angle $d^2\hat{n}'$ around a final direction $\hat{n}'$ is written as $c\kappa_S(\hat{n} \to \hat{n}'; \mathbf{x}, \nu, t)\,\rho(\mathbf{x}, t)\,d^2\hat{n}'\,dt$, where κ_S is a coefficient characterizing the scatterers, independent of the photon distribution function ℓ. In calculating the rate of change of $\ell(\hat{n}, \mathbf{x}, \nu, t)$, we must now take into account not only the scattering of photons at position $\mathbf{x}$ and time t with initial directions $\hat{n}$ into any other directions $\hat{n}'$, but also the earlier scattering of photons elsewhere with arbitrary initial directions $\hat{n}'$ into the position $\mathbf{x}$ and direction $\hat{n}$. For this purpose, we assume that $1/\kappa_S\rho$ is so much smaller than the distance over which conditions in the star vary that we can assume that any photon that after scattering reaches a given position $\mathbf{x}$ at time t can only have been scattered at a position and time where the photon distribution function ℓ

and density ρ were essentially the same as at $\mathbf{x}$ and t. (This may not be true near the surface of a star.) Then the contribution of scattering to the rate of change of ℓ is

$$\left(\frac{\partial}{\partial t}\ell(\hat{n}, \mathbf{x}, \nu, t)\right)_{\text{scattering}} = c\rho(\mathbf{x}, t)\int d^2\hat{n}' \left[-\kappa_S(\hat{n} \to \hat{n}'; \mathbf{x}, \nu, t)\ell(\hat{n}, \mathbf{x}, \nu, t)\right.$$
$$\left. + \kappa_S(\hat{n}' \to \hat{n}; \mathbf{x}, \nu, t)\ell(\hat{n}', \mathbf{x}, \nu, t)\right]. \quad (1.2.3)$$

(We are here ignoring any shift in frequency in scattering. Such shifts are small if the photon energy $h\nu$ is much less than the rest mass energy of the particles responsible for scattering, and if the velocity of these particles is much less than the speed of light, though even small frequency shifts can be important when scattering cross sections are very sensitive to frequency, as in resonant scattering.)

If (as is usually the case) the scattering is a two-body process, with photons scattered each time by a single particle of the medium, we have

$$\kappa_S(\hat{n} \to \hat{n}'; \mathbf{x}, \nu, t) = N_{\text{scat}}(\mathbf{x}, t)\sigma(\hat{n} \to \hat{n}', \nu),$$

where $\sigma(\hat{n} \to \hat{n}', \nu)$ is the differential scattering cross section, and $N_{\text{scat}}(\mathbf{x}, t)$ is the ratio of the number density of scattering centers to the mass density ρ; in other words, it is the number of scattering centers per gram.

Emission (thermal and nuclear)

We suppose that the radiation energy emitted in any direction per time, per volume, per solid angle, and per frequency interval at position $\mathbf{x}$ and time t is

$$\left(\frac{\partial}{\partial t}\ell(\hat{n}, \mathbf{x}, \nu, t)\right)_{\text{emission}} = j(\mathbf{x}, \nu, t)\rho(\mathbf{x}, t)/4\pi, \quad (1.2.4)$$

where j is another coefficient characterizing the medium and the radiation field. Note that j includes any radiation emitted isotropically subsequent to photon absorption, along with the ordinary thermal radiation from the stellar material, which is heated by nuclear processes. (Stimulated emission, which creates a photon with the same momentum and helicity as one already present, will be included as a negative term in the absorption coefficient κ_{abs}.)

Putting together these four terms, we have

$$\frac{\partial}{\partial t}\ell(\hat{n}, \mathbf{x}, \nu, t) = -c\hat{n}\cdot\nabla\ell(\hat{n}, \mathbf{x}, \nu, t)$$
$$- c\kappa_{\text{abs}}(\mathbf{x}, \nu, t)\rho(\mathbf{x}, t)\ell(\hat{n}, \mathbf{x}, \nu, t)$$
$$+ c\rho(\mathbf{x}, t)\int d^2\hat{n}'\left[-\kappa_S(\hat{n} \to \hat{n}'; \mathbf{x}, \nu, t)\ell(\hat{n}, \mathbf{x}, \nu, t)\right.$$
$$\left. + \kappa_S(\hat{n}' \to \hat{n}; \mathbf{x}, \nu, t)\ell(\hat{n}', \mathbf{x}, \nu, t)\right]$$
$$+ j(\mathbf{x}, \nu, t)\rho(\mathbf{x}, t)/4\pi. \quad (1.2.5)$$

If we now require the photon distribution function ℓ and the stellar material to be unchanging, we arrive at the condition of radiative equilibrium

$$\begin{aligned}
0 = &-c\hat{n} \cdot \nabla \ell(\hat{n}, \mathbf{x}, \nu) \\
&- c\kappa_{\text{abs}}(\mathbf{x}, \nu)\rho(\mathbf{x})\ell(\hat{n}, \mathbf{x}, \nu) \\
&+ c\rho(\mathbf{x}) \int d^2\hat{n}' \left[-\kappa_S(\hat{n} \to \hat{n}'; \mathbf{x}, \nu)\ell(\hat{n}, \mathbf{x}, \nu) \right. \\
&\left. \qquad\qquad\qquad + \kappa_S(\hat{n}' \to \hat{n}; \mathbf{x}, \nu)\ell(\hat{n}', \mathbf{x}, \nu) \right] \\
&+ j(\mathbf{x}, \nu)\rho(\mathbf{x})/4\pi,
\end{aligned} \quad (1.2.6)$$

in which we assume that κ, j, and ρ as well as ℓ are all independent of time, and so drop the argument t everywhere.

We want to use this result to derive relations between three fundamental quantities, the radiation energy per volume and per frequency interval

$$\mathcal{E}_{\text{rad}}(\mathbf{x}, \nu) \equiv \int d^2\hat{n} \, \ell(\hat{n}, \mathbf{x}, \nu), \quad (1.2.7)$$

the flux vector of radiation energy per frequency interval

$$\Phi_i(\mathbf{x}, \nu) \equiv c \int d^2\hat{n} \, \hat{n}_i \ell(\hat{n}, \mathbf{x}, \nu), \quad (1.2.8)$$

and the spatial part of the energy-momentum tensor of radiation per frequency interval

$$\Theta_{ij}(\mathbf{x}, \nu) \equiv \int d^2\hat{n} \, \hat{n}_i \hat{n}_j \ell(\hat{n}, \mathbf{x}, \nu). \quad (1.2.9)$$

(Here i and j etc. run over the Cartesian coordinate indices 1, 2, 3. Note that $\Phi_i \mathcal{N}_i \, dA \, d\nu$ is the rate at which radiant energy of frequency between ν and $\nu + d\nu$ passes through a small patch with area dA and unit normal $\mathcal{N}_i$.)

To derive our relations, we first integrate Eq. (1.2.6) over the direction of $\hat{n}$, which gives

$$\nabla \cdot \boldsymbol{\Phi}(\mathbf{x}, \nu) = -c\kappa_{\text{abs}}(\mathbf{x}, \nu)\rho(\mathbf{x})\mathcal{E}_{\text{rad}}(\mathbf{x}, \nu) + j(\mathbf{x}, \nu)\rho(\mathbf{x}). \quad (1.2.10)$$

Note that the scattering term in Eq. (1.2.6) does not contribute here, because the integrand in this term is antisymmetric in $\hat{n}$ and $\hat{n}'$.

Let us pause at this point to note a relation between the quantities $\kappa(\mathbf{x}, \nu)$, $j(\mathbf{x}, \nu)$, and $\mathcal{E}_{\text{rad}}(\mathbf{x}, \nu)$. These quantities depend only on ν and on the density $\rho(\mathbf{x})$, temperature $T(\mathbf{x})$, and chemical composition at $\mathbf{x}$; they vary with position because $\rho(\mathbf{x})$ and $T(\mathbf{x})$ and perhaps the chemical composition vary with position, but they have no independent dependence on position. That is, we can write $\kappa(\mathbf{x}, \nu)$, $j(\mathbf{x}, \nu)$, and $\mathcal{E}_{\text{rad}}(\mathbf{x}, \nu)$ as ν-dependent functions only of $\rho(\mathbf{x})$, $T(\mathbf{x})$, and chemical composition at $\mathbf{x}$. Now, if the energy emission density $j(\mathbf{x}, \nu)$ received no contribution from nuclear processes then the medium could come to equilibrium with thermal emission balancing absorption at each point and at each frequency, as in a black-body cavity. We could thus imagine a homogeneous medium that everywhere had the same temperature, density, and chemical composition that the real star has at a given position **x**. For this hypothetical homogeneous medium, Eq. (1.2.10) would require that $j = c\kappa_{abs}\mathcal{E}_{rad}$. Hence in the inhomogeneous real star, we have

$$j(\mathbf{x}, \nu) = c\kappa_{abs}(\mathbf{x}, \nu)\mathcal{E}_{rad}(\mathbf{x}, \nu) + \epsilon(\mathbf{x}, \nu), \quad (1.2.11)$$

where $\epsilon(\mathbf{x}, \nu)$ is the rate per gram and per frequency interval of energy generation from nuclear reactions. Equation (1.2.10) then reads

$$\nabla \cdot \mathbf{\Phi}(\mathbf{x}, \nu) = \epsilon(\mathbf{x}, \nu)\rho(\mathbf{x}). \quad (1.2.12)$$

We next multiply Eq. (1.2.6) with $\hat{n}_i$ and then integrate the product over the directions of $\hat{n}$:

$$\nabla_j \Theta_{ij}(\mathbf{x}, \nu) = -\kappa_{abs}(\mathbf{x}, \nu)\rho(\mathbf{x})\Phi_i(\mathbf{x}, \nu)$$
$$-c\rho(\mathbf{x}) \int d^2\hat{n}' \int d^2\hat{n}\, \hat{n}_i \left[\kappa_S(\hat{n} \to \hat{n}'; \mathbf{x}, \nu)\ell(\hat{n}, \mathbf{x}, \nu) \right.$$
$$\left. -\kappa_S(\hat{n}' \to \hat{n}; \mathbf{x}, \nu)\ell(\hat{n}', \mathbf{x})\right].$$

(In accord with the usual summation convention, the index j is here summed over the values 1, 2, 3. The emission term in Eq. (1.2.6) does not contribute here, because $j\rho$ is independent of photon direction.) Under the assumption that κ_S is invariant under rotations together of both initial and final photon directions, we may define

$$\int d^2\hat{n}'\, \kappa_S(\hat{n} \to \hat{n}'; \mathbf{x}, \nu) \equiv \kappa_{out}(\mathbf{x}, \nu) \quad (1.2.13)$$

and

$$\int d^2\hat{n}\, \hat{n}_i\kappa_S(\hat{n}' \to \hat{n}; \mathbf{x}, \nu) \equiv \hat{n}'_i\kappa_{in}(\mathbf{x}, \nu). \quad (1.2.14)$$

It follows then that

$$c \int d^2\hat{n}' \int d^2\hat{n}\, \hat{n}_i\kappa_S(\hat{n} \to \hat{n}'; \mathbf{x}, \nu)\ell(\hat{n}, \mathbf{x}, \nu) = \kappa_{out}(\mathbf{x}, \nu)\Phi_i(\mathbf{x}, \nu)$$

and

$$c \int d^2\hat{n}' \int d^2\hat{n}\, \hat{n}_i\kappa_S(\hat{n}' \to \hat{n}; \mathbf{x}, \nu)\ell(\hat{n}', \mathbf{x}, \nu) = \kappa_{in}(\mathbf{x}, \nu)\Phi_i(\mathbf{x}, \nu),$$

and therefore

$$c\nabla_j\Theta_{ij}(\mathbf{x}, \nu) = -\kappa(\mathbf{x}, \nu)\rho(\mathbf{x})\Phi_i(\mathbf{x}, \nu), \quad (1.2.15)$$

where κ is the *total opacity*:

$$\kappa(\mathbf{x}, \nu) \equiv \kappa_{abs}(\mathbf{x}, \nu) + \kappa_{out}(\mathbf{x}, \nu) - \kappa_{in}(\mathbf{x}, \nu). \quad (1.2.16)$$

To derive a formula for κ_{in} that clarifies its relation to κ_{out}, we contract Eq. (1.2.14) with $\hat{n}'$. This gives

$$\kappa_{\text{in}}(\mathbf{x}, \nu) = \int d^2\hat{n}\ (\hat{n} \cdot \hat{n}')\kappa_S(\hat{n}' \to \hat{n}; \mathbf{x}, \nu) = \int d^2\hat{n}'\ (\hat{n} \cdot \hat{n}')\kappa_S(\hat{n} \to \hat{n}'; \mathbf{x}), \tag{1.2.17}$$

which differs from the definition (1.2.13) of κ_{out} by the factor $\hat{n} \cdot \hat{n}'$. Textbook treatments of opacity often do not distinguish between absorption and scattering, and so do not encounter the term κ_{in}. This is obviously wrong, because κ_{out} would not vanish even if the scattering were restricted to an infinitesimal neighborhood of the forward direction $\hat{n}' = \hat{n}$, in which case the scattering should have no effect. The inclusion of κ_{in} removes this paradox, since

$$\kappa_{\text{out}}(\mathbf{x}, \nu) - \kappa_{\text{in}}(\mathbf{x}, \nu) = \int d^2\hat{n}'\ [1 - \hat{n} \cdot \hat{n}']\kappa_S(\hat{n} \to \hat{n}'; \mathbf{x}, \nu), \tag{1.2.18}$$

which vanishes for purely forward scattering, as it must. The authors of these treatments can get away with this oversight, because, for reasons described in Section 1.4, κ_{in} happens to vanish for Thomson scattering. But κ_{in} might matter in other scattering, such as bound–bound transitions in which the excited state decays radiatively, with the final photon direction correlated with that of the incoming photon.

So far, this has been exact, aside from the approximations made in deriving Eq. (1.2.3). We will now extend the approximation of short mean free path used there to the rest of our analysis. That is, we assume again that the opacity κ is so large that the mean path $1/\kappa\rho$ of typical photons is much smaller than the distance over which conditions vary. This is appropriate for the interiors of most stars, though not necessarily for their outer layers. It follows that to a good approximation $\ell(\hat{n}, \mathbf{x}, \nu)$ is independent of the photon direction $\hat{n}$, so that Θ_{ij} is approximately proportional to δ_{ij}. From the trace of Eq. (1.2.9) we have then

$$\Theta_{ij}(\mathbf{x}, \nu) \simeq \frac{1}{3}\delta_{ij}\mathcal{E}_{\text{rad}}(\mathbf{x}, \nu). \tag{1.2.19}$$

We also note that with $1/\kappa\rho$ very short the radiation is in thermal equilibrium with local matter at a temperature T, so that

$$\mathcal{E}_{\text{rad}}(\mathbf{x}, \nu) \simeq B(\nu, T(\mathbf{x})), \tag{1.2.20}$$

where B is the Planck black-body distribution

$$B(\nu, T) = \frac{8\pi h}{c^3} \frac{\nu^3}{\exp(h\nu/k_B T) - 1}. \tag{1.2.21}$$

Using Eqs. (1.2.19) and (1.2.20) in Eq. (1.2.15),

$$c\,\nabla B(\nu, T(\mathbf{x})) = -3\kappa(\mathbf{x}, \nu)\rho(\mathbf{x})\mathbf{\Phi}(\mathbf{x}, \nu). \tag{1.2.22}$$

1.2 Radiative Energy Transport

Of course, $\ell(\hat{n}, \mathbf{x}, \nu)$ does depend somewhat on $\hat{n}$. Even deep in a star, there is some difference between up and down, the directions toward and away from the star's surface. We are neglecting this in Eqs. (1.2.19) and (1.2.20), but since $\kappa\rho$ is assumed large, we may not neglect the quantity $\kappa\rho\Phi_i$ in Eq. (1.2.22), even though perfect isotropy of the photon distribution would make Φ_i vanish.

Now let us take up the special case of greatest interest, a spherically symmetric star in which the only special direction at any point is the radial direction, which distinguishes up and down. We then take the flux vector to point in the direction $\hat{x} \equiv \mathbf{x}/r$, and otherwise to depend only on ν and $r \equiv |\mathbf{x}|$, so that we may write

$$\Phi(\mathbf{x}, \nu) = \hat{x}\,\frac{\mathcal{L}(r, \nu)}{4\pi r^2}. \qquad (1.2.23)$$

Then $\mathcal{L}(r, \nu)$ is the total radiant energy flux, the radiant energy per time and per frequency interval passing outward through a sphere of radius r. In this case, Eqs. (1.2.12) and (1.2.22) take the form

$$\frac{d\mathcal{L}(r, \nu)}{dr} = 4\pi r^2 \epsilon(r, \nu)\rho(r), \qquad (1.2.24)$$

and

$$c\,\frac{d\,B(\nu, T(r))}{dr} = -3\kappa(r, \nu)\rho(r)\frac{\mathcal{L}(r, \nu)}{4\pi r^2}. \qquad (1.2.25)$$

To calculate the temperature distribution in a star, it suffices to consider the total radiant energy for all frequencies. The total radiant energy flux is defined by

$$\mathcal{L}(r) \equiv \int d\nu\, \mathcal{L}(r, \nu), \qquad (1.2.26)$$

and the total energy per gram emitted by nuclear processes at all frequencies is

$$\epsilon(r) \equiv \int d\nu\, \epsilon(r, \nu). \qquad (1.2.27)$$

Then integrating Eq. (1.2.24) over frequency, we have

$$\frac{d\mathcal{L}(r)}{dr} = 4\pi r^2 \epsilon(r)\rho(r). \qquad (1.2.28)$$

In order to write the equation for dT/dr in terms of $\mathcal{L}(r)$, we divide Eq. (1.2.25) by $\kappa(r, \nu)$ and integrate over ν:

$$-3\rho(r)\frac{\mathcal{L}(r)}{4\pi r^2} = c\int d\nu\,\frac{1}{\kappa(r, \nu)}\left(\frac{\partial B(\nu, T)}{\partial T}\right)_{T=T(r)} T'(r).$$

We define the *Rosseland mean opacity*[4] $\kappa(r)$ as the inverse of the average of the inverse of $\kappa(r, \nu)$, evaluated with a weighting function $(\partial B(\nu, T)/\partial T)_{T=T(r)}$:

$$\int d\nu \, \frac{1}{\kappa(r,\nu)} \left(\frac{\partial B(\nu,T)}{\partial T}\right)_{T=T(r)} \equiv \frac{1}{\kappa(r)} \int d\nu \left(\frac{\partial B(\nu,T)}{\partial T}\right)_{T=T(r)} = \frac{4aT^3(r)}{\kappa(r)}, \quad (1.2.29)$$

where a is the radiation energy constant, $a = 8\pi^5 k_B^4/15h^3c^3 = 7.566 \times 10^{-15}$ erg cm^{-3} K^{-4}. So

$$-3\rho(r)\frac{\mathcal{L}(r)}{4\pi r^2} = \frac{4acT^3(r)T'(r)}{\kappa(r)},$$

or, multiplying by $\kappa(r)/4acT^3(r)$:

$$\frac{dT(r)}{dr} = -\frac{3\rho(r)\kappa(r)}{4acT^3(r)} \frac{\mathcal{L}(r)}{4\pi r^2}. \quad (1.2.30)$$

Equations (1.2.28) and (1.2.30) are the fundamental equations of radiative energy transport in spherical star interiors.

It is convenient for some purposes to introduce an opacity function $\kappa(\rho, T, \nu)$ and its Rosseland mean $\kappa(\rho, T)$ that depend on density and temperature rather than on position, with

$$\kappa(r) = \kappa(\rho(r), T(r)), \quad \kappa(r, \nu) = \kappa(\rho(r), T(r), \nu). \quad (1.2.31)$$

Then the definition (1.2.29) of the Rosseland mean takes the position-independent form

$$\int d\nu \, \frac{1}{\kappa(\rho, T, \nu)} \left(\frac{\partial B(\nu, T)}{\partial T}\right) = \frac{4aT^3}{\kappa(\rho, T)}. \quad (1.2.32)$$

1.3 Radiative Models

In this section we shall describe the differential equations and boundary conditions that govern a star in which energy transport is everywhere dominated by radiation. The most important result here is that for a set of stars of a given age and initial uniform chemical composition (such as the stars in many clusters), any stellar parameter, such as radius, luminosity, etc., may be expressed as a function of stellar mass. In consequence, when any two of these parameters are plotted against one another, the plot is a one-dimensional curve. (One such relation is the plot of luminosity against effective temperature, known as the Hertzsprung–Russell relation, about which more later.) The following two

---
[4] S. Rosseland, *Mon. Not. Roy. Astron. Soc.* **84**, 525 (1924).

sections will consider the opacity and nuclear energy generation per mass, which appear as ingredients in these differential equations. Then in Section 1.6 we will derive consequences from these equations in the form of power laws for various stellar properties for stars that are on the main sequence of the Hertzsprung–Russell diagram. Section 1.7 considers energy transport by convection, and shows that convection does not affect the main results of this section and Section 1.6.

With the chemical composition fixed and uniform, we can regard the pressure $p(r)$, opacity $\kappa(r)$, and nuclear energy production per mass $\epsilon(r)$ as fixed functions of the density $\rho(r)$ and temperature $T(r)$. The star's structure is then described by four functions of the radial coordinate r: the mass $\mathcal{M}(r)$ contained within a sphere of radius r; the radiant energy per second $\mathcal{L}(r)$ flowing outward through a spherical surface of radius r; and the density $\rho(r)$ and temperature $T(r)$. These four quantities are governed by four first-order differential equations: the equations (1.1.4) and (1.1.5) of hydrostatic equilibrium

$$\frac{dp(r)}{dr} = -\frac{G\mathcal{M}(r)\rho(r)}{r^2} \tag{1.3.1}$$

and

$$\frac{d\mathcal{M}(r)}{dr} = 4\pi r^2 \rho(r), \tag{1.3.2}$$

and the equations (1.2.28) and (1.2.30) of radiative energy transport

$$\frac{d\mathcal{L}(r)}{dr} = 4\pi r^2 \epsilon(r)\rho(r) \tag{1.3.3}$$

and

$$\frac{dT(r)}{dr} = -\frac{3\kappa(r)\rho(r)}{4caT^3(r)}\frac{\mathcal{L}(r)}{4\pi r^2}. \tag{1.3.4}$$

There are also four boundary conditions – two at the center,

$$\mathcal{M}(0) = \mathcal{L}(0) = 0; \tag{1.3.5}$$

and two at the star's nominal radius R,

$$\rho(R) = T(R) = 0. \tag{1.3.6}$$

With the pressure p, Rosseland mean opacity κ, and nuclear energy production per mass ϵ assumed to be given as functions of density and temperature, the differential equations (1.3.1)–(1.3.4) and boundary conditions (1.3.5) and (1.3.6) then govern the four unknown functions $\rho(r)$, $\mathcal{M}(r)$, $T(r)$, and $\mathcal{L}(r)$.

Before considering the implications of these differential equations and boundary conditions, we need to say a bit about the implausible boundary condition that the temperature and density vanish at the star's surface. With four first-order differential equations for four unknown functions, and only two boundary conditions at $r = 0$, there is enough freedom to impose these two additional conditions at any radius in the generic case. We call the value of r where these conditions are imposed on the solutions of Eqs. (1.3.1)–(1.3.4) the "nominal radius" R of the star. But of course the surfaces of stars are not actually at absolute zero temperature. Not even close. In fact, the approximation of nearly perfect isotropy that we used in deriving the equations (1.3.3) and (1.3.4) breaks down close to the stellar surface, where there is a big difference between up, down, and sideways. Specifically, this approximation breaks down at values of r for which $R - r$ is no longer large compared with the typical photon free path $1/\rho(r)\kappa(r)$ at r. In this region, known as the stellar atmosphere, we need to use the full equation (1.2.6) of radiative equilibrium, and we do not find a surface with absolute zero temperature. The nominal radius R is where the density and temperature *would* vanish if Eqs. (1.3.1)–(1.3.4) held out to this radius.

In the real world, instead of a surface at which the density and temperature vanish, there is a "true surface" with radius R_{true} beyond which there is essentially empty space, with only outgoing radiation and some gas of very low density, such as the solar corona. But this is not the surface from which comes the light we see. To the extent that the light of a star resembles black-body radiation, we can think of it as coming from an effective surface with radius R_{eff}, defined by the condition

$$\sigma T^4(R_{\text{eff}}) \times 4\pi R_{\text{eff}}^2 = L, \tag{1.3.7}$$

where $\sigma = ac/4$ is the Stefan–Boltzmann constant, and L is the star's luminosity, the value of $\mathcal{L}(r)$ at all values of r outside the stellar core in which nuclear energy production occurs. The depth of the effective surface below the true surface is best described in terms of its optical depth

$$\tau_{\text{eff}} = \int_{R_{\text{eff}}}^{R_{\text{true}}} \kappa(r)\rho(r)\,dr. \tag{1.3.8}$$

Since it is the typical photon free path $1/\kappa\rho$ that sets the scale of variations with radius near the surface, we expect τ_{eff} to be of order unity. (In fact, there is a time-honored but rather unconvincing calculation[5] that gives the optical depth of the effective surface as $\tau_{\text{eff}} = 2/3$.)

The important point for us is that the thickness of the stellar atmosphere is much less than R. As long as we restrict our interest to the star's interior, we can therefore continue to use the differential equations (1.3.1)–(1.3.4), with the boundary conditions (1.3.5) and (1.3.6), with the understanding that the condition (1.3.6) just means that the density and pressure are much less at the star's true surface than deep in the interior. For instance, the central density and temperature of the Sun are (98 ± 15) g/cm^3 and $(13.6 \pm 1.2) \times 10^6$ K, while even deep in the stellar atmosphere, at an optical depth $\tau = 10$, the solar density and

---

[5] For instance, see J. P. Cox and R. T. Giuli, *Principles of Stellar Structure: Application to Stars*, Vol. 2 (Gordon & Breach, New York, 1968), Chapter 20.

temperature are only about 5×10^{-7} g/cm^3 and 9,700 K, much less than the central values.

With four first-order equations and four boundary conditions in which there appear only a single parameter R, we expect a one-parameter family of solutions. This result is close to a conclusion that is often called the *Vogt–Russell theorem*,[6] which asserts that for a definite chemical composition there is a unique solution to the equations of stellar structure, that depends on just a single stellar parameter, such as the radius R or the total mass M. In fact, we can't be sure of the existence of a solution, because it is possible that a singularity could be encountered that prevents a solution, though no such case of astronomical relevance is known. Also, assuming a solution exists, it may not be unique.

The possibility of non-uniqueness arises from the peculiar feature, that the boundary conditions refer to two different boundaries, $r = 0$ and $r = R$. Consider how we would actually construct a solution. Starting at $r = 0$, we can adopt various trial values ρ_c and T_c of the central density $\rho(0)$ and central temperature $T(0)$, so that with the original conditions $\mathcal{M}(0) = \mathcal{L}(0) = 0$ we have four initial conditions. Integrating Eqs. (1.3.1)–(1.3.4) with these initial conditions gives a unique solution, depending on ρ_c and T_c. We can then adjust these two initial values so that the other conditions, $\rho(R) = T(R) = 0$, are satisfied at any given R. With two conditions on the two parameters ρ_c and T_c, there is likely to be a solution, but possibly more than one. As long as the number of solutions is finite, they can each depend on only a single free parameter, which so far we have taken as the stellar radius R.

Of course, if all stellar parameters depend on a single parameter R, they can be taken to depend on any one of the other stellar parameters, not necessarily R. In particular, since the stellar mass M is the one thing that remains essentially fixed as a star evolves (until the star in its old age begins to blow off mass), it is more natural to take the single parameter as M rather than R. We can (though we need not) do this directly, by a reinterpretation of the differential equations. We can take the independent variable to be $\mathcal{M}$ rather than r, with the dependent variables taken as $r(\mathcal{M})$ along with $\rho(\mathcal{M})$, $T(\mathcal{M})$, and $\mathcal{L}(\mathcal{M})$. The differential equations are the reciprocal of Eq. (1.3.2),

$$\frac{dr(\mathcal{M})}{d\mathcal{M}} = \frac{1}{4\pi r^2(\mathcal{M})\rho(\mathcal{M})}, \qquad (1.3.9)$$

and the ratios of Eqs. (1.3.1), (1.3.3), and (1.3.4) to Eq. (1.3.2):

$$\frac{dp(\mathcal{M})}{d\mathcal{M}} = -\frac{G\mathcal{M}}{4\pi r^4(\mathcal{M})}, \qquad (1.3.10)$$

$$\frac{d\mathcal{L}(\mathcal{M})}{d\mathcal{M}} = \epsilon(\mathcal{M}), \qquad (1.3.11)$$

---

[6] H. Vogt, *Astron. Nachr.* **226**, 301 (1926); H. N. Russell, *Astronomy* (Boston) **2**, 910 (1927).

and
$$\frac{dT(\mathcal{M})}{d\mathcal{M}} = -\frac{3\kappa(\mathcal{M})\mathcal{L}(\mathcal{M})}{4caT^3(\mathcal{M})\left(4\pi r^2(\mathcal{M})\right)^2}. \qquad (1.3.12)$$

Instead of imposing boundary conditions at $r = 0$ and $r = R$, here they are imposed at $\mathcal{M} = 0$,
$$r(\mathcal{M}) = \mathcal{L}(\mathcal{M}) = 0 \text{ at } \mathcal{M} = 0, \qquad (1.3.13)$$
and at $\mathcal{M}$ equal to the total stellar mass M,
$$\rho(\mathcal{M}) = T(\mathcal{M}) = 0 \text{ at } \mathcal{M} = M. \qquad (1.3.14)$$

With the equations written in this way, there is no need to input any stellar parameter aside from the mass M.

It is the dependence of stellar structure on just a single parameter that explains a remarkable feature of observations of clusters of stars. The dozens or hundreds of stars in an open cluster like the Pleiades generally condensed at about the same time from the same cloud of interstellar material, so they all have pretty much the same initial chemical composition and age as well as distance, though differing widely in their masses. The only thing on which any observable feature of the stars in such a cluster can depend that varies from one star to another will thus be the stars' masses. Hence when any pair of observables for the cluster stars are plotted against each other, these points will fall on a one-dimensional curve, each different point on this curve corresponding to a different stellar mass.

This is less so for the thousands or hundreds of thousands of stars in a globular cluster like M15, where there is a greater spread in age and initial chemical composition. But even here the plot of any pair of observables against each other is a more or less thickened curve.

The most easily observable stellar quantities are the luminosity L (or, if the distance d to the cluster is not known, the apparent luminosity $L/4\pi d^2$) and the effective temperature T_{eff}. The effective temperature is defined by the condition that $L = \sigma T_{\text{eff}}^4 \times 4\pi R^2$, but it is estimated from observations of the star's color[7] and/or spectrum, as described in the following table:[8]

---

[7] The color of a star is measured by the differences of its luminosity when the star is observed with several different filters. As seen by an observer without filters, the color depends on the distribution with frequency of the radiant energy emitted by the star, for those frequencies that are visible to the eye. For hot stars with temperatures $T > 30,000$ K, these frequencies are all much less than $k_B T/h$, and therefore, according to the black-body formula (1.2.21), the energy emitted between visible frequencies ν and $\nu + d\nu$ is proportional to $\nu^2 d\nu$. As it happens, this is the same frequency distribution as for the light scattered by molecules and other small particles in the atmosphere, which gives the sky its color. Hence sky blue is the asymptotic visible color of black bodies with very high temperature.

[8] The information here is taken from F. LeBlanc, *Introduction to Stellar Atmospheres* (John Wiley & Sons, Chichester, 2010), with some additions from other sources.

1.3 Radiative Models

Typical spectral lines, effective temperatures, colors, and examples of various types of star

Type	Lines	T_{eff} (K)	Color	Example
O	HeII abs	>30,000	Sky blue	λ Ori
B	HeI abs, H	10,000–30,000	Blue–White	Rigel
A	H, CaII	7,500–10,000	White	Sirius A, Vega
F	CaII, H weaker	6,000–7,500	Yellow–White	Procyon
G	CaII, Fe, H weak	5,000–6,000	Yellow	Sun
K	Metals, CH, CN	3,500–5,000	Orange	Arcturus
M	TiO	<3,500	Red	Antares

The graph of observed absolute or apparent luminosity versus effective temperature is known as the Hertzsprung–Russell diagram, which was first constructed a century ago.[9]

In practice, the Hertzsprung–Russell diagram of a cluster is a thick curve, not strictly one-dimensional. This is because the cluster stars did not all begin at precisely the same time with precisely the same chemical composition. There are also observational problems: a star's color and spectrum do not give a precise value for the effective temperature, and it is often difficult to distinguish binary stars from single stars. Even so, one can clearly see in the data that there is a one-dimensional curve of luminosity versus effective temperature, not just points everywhere in the plot.

The Hertzsprung–Russell diagram for a cluster commonly contains a *main sequence*, consisting of stars like the Sun that are still burning hydrogen at their cores. On the main sequence L increases smoothly with T_{eff}, with the most massive stars the hottest and most luminous. (In Section 1.6 we will show how to estimate the shape of the main sequence curve by applying dimensional analysis to Eqs. (1.3.1)–(1.3.4).) As the cluster evolves, the Hertzsprung–Russell diagram develops a red giant branch, consisting of stars that have converted most of the hydrogen at their cores to helium, and are burning hydrogen only in a shell around the inert helium core. On this branch, the effective temperature *decreases* (and radius increases) with increasing luminosity, accounting for the red color of very luminous red giant stars such as Betelgeuse and Antares. The heavier stars on the main sequence have larger L and therefore evolve more quickly, so as time passes more and more of the upper part of the main sequence bends over into the red giant branch. Observations of this main sequence

---
[9] E. Hertzsprung, *Astron. Nachr.* **179** (24), 373 (1908); H. N. Russell, *Pop. Astron.* **22**, 275 (1914).

turn-off therefore indicate the age of the cluster.[10] Eventually the more massive stars of the cluster will begin to burn helium, and the Hertzsprung–Russell diagram will develop further complications, but it remains a more-or-less one-dimensional curve, as required by the Vogt–Russell theorem.

There is a general conclusion of some importance, which can be derived immediately from Eqs. (1.3.1)–(1.3.4), without detailed calculation. We note that the pressure p in Eq. (1.3.1) is the sum of the pressures of gas and radiation,

$$p = p_{\text{gas}} + p_{\text{rad}}, \qquad (1.3.15)$$

where, for black-body radiation,

$$p_{\text{rad}} = \frac{a}{3} T^4. \qquad (1.3.16)$$

For an ideal gas $p_{\text{gas}} = \rho k_B T / m_1 \mu$, where μ is the molecular weight and m_1 is the nucleon mass, or more precisely, the mass of unit atomic weight. For the present all we need to know about the gas pressure is that it decreases with increasing r. Now, Eq. (1.3.4) may be written

$$\frac{dp_{\text{rad}}(r)}{dr} = -\frac{\kappa(r) \rho(r) \mathcal{L}(r)}{4\pi c r^2}.$$

Taking the difference between this and Eq. (1.3.1) gives

$$-\frac{\kappa(r) \rho(r) \mathcal{L}(r)}{4\pi c r^2} + \frac{G \mathcal{M}(r) \rho(r)}{r^2} = -\frac{dp_{\text{gas}}(r)}{dr} > 0$$

and therefore, everywhere in the star,

$$\kappa(r) \mathcal{L}(r) < 4\pi G c \mathcal{M}(r).$$

In particular, by setting r equal to the nominal stellar radius R, we find an inequality involving the star's luminosity $L = \mathcal{L}(R)$ and mass $M = \mathcal{M}(R)$:

$$\kappa(R) L < 4\pi G c M. \qquad (1.3.17)$$

If this inequality were violated, then the radiation pressure alone would be strong enough to blow off the outer layers of the star. In the commonly encountered case where the opacity in the star's outer layers is due to Thomson scattering the inequality (1.3.17) is known as the *Eddington limit*. This inequality also limits the luminosity that can be produced by spherically symmetric accretion onto a star or galactic nucleus.

This derivation also shows that if gas pressure were negligible compared with radiation pressure (as it is in only the most massive stars) the inequality would become an equality, $\kappa(R) L = 4\pi G c M$.

---

[10] For a summary of the use of this technique in cosmology, see S. Weinberg, *Cosmology* (Oxford University Press, Oxford, 2008), pp. 62–63.

1.4 Opacity

We saw in Section 1.2 that Eq. (1.2.30), one of the pair of equations that govern the variation of temperature of stars with distance r from the center, involves a quantity $\kappa(r)$, known as the opacity. In general, the opacity is given by Eq. (1.2.16):

$$\kappa \equiv \kappa_{\text{abs}} + \kappa_{\text{out}} - \kappa_{\text{in}}, \qquad (1.4.1)$$

with it understood that in Eq. (1.2.30) $\kappa(r)$ is a Rosseland mean value $\kappa(\rho(r), T(r))$, calculated according to Eq. (1.2.32):

$$\int dv \, \frac{1}{\kappa(\rho, T, \nu)} \left(\frac{\partial B(\nu, T)}{\partial T} \right) = \frac{4aT^3}{\kappa(\rho, T)},$$

where B is the black-body distribution function

$$B(\nu, T) = \frac{8\pi h}{c^3} \frac{\nu^3}{\exp(h\nu/k_B T) - 1}.$$

The first term in Eq. (1.4.1) is defined so that $c\rho\kappa_{\text{abs}}$ is the *net* rate of absorption – that is, it is the average rate per photon at which photons are absorbed, less the rate per initial photon at which photons with the same momentum are created by stimulated emission. If Γ_{abs} is the rate of absorption alone, then when stimulated emission is taken into account, the net rate of photon absorption is

$$c\rho\kappa_{\text{abs}}(\rho, T, \nu) = \Gamma_{\text{abs}}(\rho, T, \nu)\left[1 - e^{-h\nu/k_B T}\right]. \qquad (1.4.2)$$

This can most easily be seen by returning to Eqs. (1.2.11) and (1.2.20), which show that when radiation and matter come to equilibrium in the absence of nuclear energy generation, the absorption opacity is related to the energy $j(\rho, T, \nu)$ emitted by the matter per mass, per time, and per frequency interval, by

$$\kappa_{\text{abs}}(\rho, T, \nu) = j(\rho, T, \nu)/cB(\nu, T) = \frac{c^2}{8\pi h \nu^3} j(\rho, T, \nu)\left[\exp(h\nu/k_B T) - 1\right].$$

The emission rate j has a familiar factor $\exp(-h\nu/k_B T)$, reflecting the probability of excitation by energy $h\nu$ of degrees of freedom in the matter. When combined with the factor $\exp(h\nu/k_B T) - 1$ from $1/B$ this gives the correction factor $1 - e^{-h\nu/k_B T}$ in Eq. (1.4.2), in which the first and second terms arise from absorption and stimulated emission.[11]

The second and third terms in Eq. (1.4.1) are defined so that $c\rho\kappa_{\text{out}}$ and $c\rho\kappa_{\text{in}}$ are the rates at which photons are scattered out of or into any given direction.

---

[11] For a derivation of Eq. (1.4.2) that does not depend on the assumption that the radiation can come into equilibrium with the matter, see R. Flauger and S. Weinberg, *Phys. Rev. D* **99**, 123030 (2019).

In cases where scattering occurs in a collision with a single particle, such as an electron or atom, these terms are given by Eqs. (1.2.13) and (1.2.17):

$$\kappa_{\text{out}} = N_{\text{scat}} \int d^2\hat{n}' \, \sigma_{\text{scat}}(\hat{n} \to \hat{n}'), \tag{1.4.3}$$

$$\kappa_{\text{in}} = N_{\text{scat}} \int d^2\hat{n}' \, (\hat{n}' \cdot \hat{n}) \, \sigma_{\text{scat}}(\hat{n} \to \hat{n}'), \tag{1.4.4}$$

where $\sigma_{\text{scat}}(\hat{n} \to \hat{n}')$ is the differential cross section for scattering of a photon traveling in a direction $\hat{n}$ into a direction $\hat{n}'$, and N_{scat} is the number of scatterers per gram. (These integrals are independent of the unit vector $\hat{n}$ because of the invariance of the integrands under simultaneous rotations of $\hat{n}$ and $\hat{n}'$.)

Now let us consider the various contributions to opacity, and the temperature and density dependence of each. It is often a fair approximation to represent the opacity as a simple function of temperature and density, proportional to powers of both:

$$\kappa(\rho, T) = \kappa_1 \rho^\alpha (k_B T)^\beta, \tag{1.4.5}$$

where κ_1 as well as α and β are approximately independent of density and temperature. We will estimate α and β below for contributions to opacity of various types, and show in Section 1.6 how these results can be used to relate observable properties of stars.

Thomson Scattering

This is the simplest contribution to opacity. It is the elastic scattering of photons with energies much less than $m_e c^2$ on free electrons moving non-relativistically. The differential scattering cross section is

$$\sigma_{\text{Thomson}}(\hat{n} \to \hat{n}') = \frac{e^4}{2m_e^2 c^4} \left[1 + (\hat{n} \cdot \hat{n}')^2 \right]. \tag{1.4.6}$$

(Recall that in this book e is the charge of the electron in unrationalized electrostatic units.) Because this differential cross section is even[12] in $\hat{n}'$, while the factor $\hat{n} \cdot \hat{n}'$ in Eq. (1.4.4) is odd in $\hat{n}'$, here we have $\kappa_{\text{in}} = 0$. Hence, where the opacity is dominated by Thomson scattering, the total opacity is

$$\kappa = \kappa_{\text{out}} = N_e \sigma_T, \tag{1.4.7}$$

---

[12] This forward–backward symmetry can be understood in classical terms. Classically, in Thomson scattering the electron position oscillates under the influence of the electric field of the incoming photon, and this oscillation produces the electromagnetic field of the outgoing photon. This oscillation is in the direction of the polarization vector of the incoming photon, which is normal to the photon's direction, so there is nothing about this oscillation or the field it produces that can distinguish between the forward and backward directions.

where σ_T is the total Thomson scattering cross section, given by the integral of the differential cross section (1.4.6) over solid angle:

$$\sigma_T = \frac{8\pi}{3}\left(\frac{e^2}{\hbar c}\right)^2 \left(\frac{\hbar}{m_e c}\right)^2 = 0.66525 \times 10^{-24}\,\text{cm}^2,$$

and N_e is the number of free electrons per gram. For instance, for a medium consisting of completely ionized atoms of atomic number Z and atomic weight A, we have $N_e = Z/Am_1$, where $m_1 = 1.66054 \times 10^{-24}$ g is the mass for unit atomic weight. This gives a Thomson scattering opacity (1.4.7) equal to $0.400 \times Z/A$ cm^2/g.

Since the cross section is constant (aside from a possible dependence of the degree of ionization on temperature and density) the opacity for Thomson scattering has

$$\alpha = \beta = 0. \tag{1.4.8}$$

No averaging over photon frequency is necessary if Thomson scattering dominates the opacity.

Free–Free Absorption

In the absence of external fields, the conservation of energy and momentum forbids the absorption of a photon by a free electron. If the photon has momentum $\mathbf{q}$ then it has energy $c|\mathbf{q}|$, so the conservation of energy and momentum requires that

$$0 = (E' - E)^2 - c^2(\mathbf{p}' - \mathbf{p})^2 = 2m_e^2 c^4 - 2E'E + 2c^2 \mathbf{p}' \cdot \mathbf{p}.$$

where $\mathbf{p}$ and $\mathbf{p}'$ are the initial and final electron momenta, and $E = [c^2\mathbf{p}^2 + m_e^2 c^4]^{1/2}$ and $E' = [c^2\mathbf{p}'^2 + m_e^2 c^4]^{1/2}$ are the initial and final electron energies. This is not possible if any energy is absorbed by the electron, for in the frame in which the electron is initially at rest, this requires that $E' = m_e c^2$, so the final electron would have to be also at rest in the same frame.

But in the Coulomb field of an atomic nucleus, the nucleus can take up momentum without carrying away appreciable energy because it is so massive. So absorption is possible on a free electron near a nucleus, with the energy but not the momentum of electron and photon conserved, in the same way that a dropped ball can bounce upward without losing energy, its momentum being taken up by the Earth. This is the inverse of the familiar process of *bremsstrahlung*, in which a photon is emitted when a charged particle is slowed in a collision. (The cooling of interstellar matter by bremsstrahlung is discussed at the end of Section 3.3, and the emission of detectable radiation by bremsstrahlung is considered in Section 3.7.) The absorption of photons by free electrons in the Coulomb field of a nucleus leads to what is known as *Kramers opacity*, named for Hendrik Kramers (1894–1952) who, using classical physics, first attempted a calculation.[13] Kramers' classical result was in effect that the rate of absorption of a photon of frequency ν (averaged over photon directions and helicities) is[14]

$$\Gamma_{\text{Kramers}}(\rho, T, \nu) = \int n_e(\mathbf{v}, T) d^3v \, \frac{4\pi Z^2 e^6 n_N}{3\sqrt{3} h m_e^2 v \nu^3}$$

where the integral is over initial electron velocities $\mathbf{v}$; $n_e(\mathbf{v}, T)$ is the number of electrons per spatial volume and per velocity-space volume; n_N is the number density of ions, taken to have charge Ze; e is the magnitude of the electron charge in unrationalized electrostatic units; and $h = 2\pi\hbar$.

Depending on the electron velocity and photon frequency, this can be significantly modified by quantum and other corrections. With or without these corrections, the net rate $c\rho\kappa$ of photon absorption in free–free transitions is quadratic in particle densities, so $\alpha = 1$, but the temperature dependence is more complicated. It was first calculated by John Arthur Gaunt[15] (1904–1944). It has become traditional to express the rate per electron as the Kramers result multiplied by a correction factor, known as the free–free Gaunt factor:

$$\Gamma_{\text{ff abs}}(\rho, T, \nu) = \int n_e(\mathbf{v}, T) d^3v \, \frac{4\pi Z^2 e^6 n_N}{3\sqrt{3} h m_e^2 v \nu^3} g_{\text{ff}}(\nu, v). \qquad (1.4.9)$$

This absorption rate is quite complicated, given by an integral of the matrix element of the momentum operator of the electron between initial and final electron wave functions, which in a Coulomb potential are Kummer functions. But it is not so difficult to carry out the calculation in Born approximation – that is, to first order in the Coulomb potential. As shown in the appendix to this section, in this order the rate at which a photon of frequency ν is absorbed is[16]

$$\Gamma_{\text{ff abs}}(\rho, T, \nu) = \int n_N n_e(\mathbf{v}, T) d^3v \, \frac{4 Z^2 e^6}{3 h m_e^2 v \nu^3} \ln\left(\frac{v' + v}{v' - v}\right), \qquad (1.4.10)$$

where v' is the final electron velocity, given by the energy conservation condition

$$\frac{m_e v'^2}{2} = \frac{m_e v^2}{2} + h\nu. \qquad (1.4.11)$$

---

[13] H. Kramers, *Phil. Mag.* **46**, 836 (1923).
[14] The fractional rate of decrease of energy in a light ray of frequency ν is $h\nu\Gamma(\nu)$, which for the Kramers formula is independent of Planck's constant. It is this rate that emerges from a purely classical calculation.
[15] J. A. Gaunt, *Proc. Roy. Soc.* **126**, 654 (1930).
[16] For a different derivation of this formula, using "old-fashioned" second-order perturbation theory, see H.-Y. Chiu, *Stellar Physics* (Blaisdell, Waltham, MA, 1968). The factor v in the denominator of Eq. (1.4.10) appears in Chiu's book as v'; presumably this is a typographical error.

1.4 Opacity

That is, the Gaunt factor is

$$g_{\text{ff}}(v', v) = \frac{\sqrt{3}}{\pi} \ln\left(\frac{v' + v}{v' - v}\right), \quad (1.4.12)$$

with v' again given by Eq. (1.4.11). This is a good approximation for non-relativistic electrons if the Coulomb potential at an electron scattered by an atom or ion is typically much less than electron kinetic energies, which is the case if $Ze^2/\hbar v \ll 1$ and $Ze^2/\hbar v' \ll 1$.

In thermal equilibrium at temperature T, far from degeneracy, the electron velocity distribution is given by the Maxwell–Boltzmann formula

$$n_e(\mathbf{v}, T) = n_e \left(\frac{m_e}{2\pi k_B T}\right)^{3/2} \exp\left(-\frac{m_e v^2}{2k_B T}\right), \quad (1.4.13)$$

where n_e is the total electron number density. We can find the temperature dependence of the integral (1.4.10) by introducing a re-scaled variable of integration

$$x \equiv v\sqrt{m_e/2k_B T}.$$

Then Eq. (1.4.10) can be written

$$\Gamma_{\text{ff abs}}(\rho, T, \nu) = n_e n_N \frac{16 Z^2 e^6}{3hc m_e^2 v^3} \sqrt{\frac{m_e}{2\pi k_B T}} \int_0^\infty x e^{-x^2} dx \times \ln\left(\frac{x' + x}{x' - x}\right), \quad (1.4.14)$$

where n_e and n_N are the total number densities of electrons and ions, respectively. If we supply the correction factor $1 - \exp(-h\nu/k_B T)$ for stimulated emission, and as usual write the result as $c\rho\kappa_{\text{ff}}$, then

$$\kappa_{\text{ff}}(\rho, T, \nu) = \rho N_e N_N \frac{16 Z^2 e^6}{3hc m_e^2 v^3} \sqrt{\frac{m_e}{2\pi k_B T}} \int_0^\infty x e^{-x^2} dx$$

$$\times \ln\left(\frac{x' + x}{x' - x}\right) \left(1 - \exp(-h\nu/k_B T)\right), \quad (1.4.15)$$

where $N_e \equiv n_e/\rho$ is the number of electrons per gram, $N_N \equiv n_N/\rho$ is the number of nuclei per gram, and $x' \equiv v'\sqrt{m_e/2k_B T}$ is given by the energy conservation equation (1.4.11) as

$$x'^2 = x^2 + y, \quad y \equiv h\nu/k_B T. \quad (1.4.16)$$

The Rosseland mean opacity (1.2.32) is here

$$\kappa(\rho, T) = \frac{8\rho(k_B T)^{-7/2} N_e N_N Z^2 e^6 h^6 (a/k_B^4) m_e^{-3/2}}{3\sqrt{2}\pi^{3/2} \int_0^\infty dy \frac{y^6 e^y}{(e^y - 1)} \left[\int_0^\infty x e^{-x^2} dx \times \ln\left(\frac{x' + x}{x' - x}\right)\right]^{-1}}, \quad (1.4.17)$$

with x' related to the integration variables x and y by the energy conservation condition (1.4.16). The important result is that in Eq. (1.4.5) the Kramers opacity has

$$\alpha = 1, \quad \beta = -7/2. \tag{1.4.18}$$

The mean opacity has a factor $T^{-7/2}$ because of the factor $1/\sqrt{T}$ in Eq. (1.4.15), and because the factor $1/\nu^3$ in Eq. (1.4.15) is converted into a factor proportional to $1/T^3$ in the Rosseland mean.

It should not be thought that the $T^{-7/2}$ dependence of the free–free opacity continues to arbitrary low temperatures. Obviously, for sufficiently low temperatures, there are very few free electrons, and the free–free and Thomson scattering contributions to the opacity both become negligible.

High-Energy Bound–Free Absorption

When a photon is absorbed by a bound electron whose binding energy is much less than the photon energy, it hardly matters that the electron is initially bound. Thus the temperature dependence in this case is the same as for free–free absorption, with $\beta = -7/2$. The difference is that the relevant density of electrons is not the ambient density of free electrons, but an average square of the bound electron wave function, so the absorption rate $c\rho\kappa$ is proportional just to the density of atoms, and hence $\alpha = 0$ rather than $\alpha = 1$. The contribution to opacity of this sort of photon absorption is often lumped in with free–free absorption in what is called Kramers opacity.

Bound–Bound Absorption and Low-Energy Bound–Free Absorption

In these cases the photon is absorbed by a bound electron whose binding energy is at least comparable to the photon energy. This contribution to opacity involves complications of atomic physics not present for other contributions, and will not be examined further here. The heating of interstellar hydrogen by low-energy bound–free absorption of photons from hot stars is discussed in Section 3.2.

Appendix: Calculation of Free–Free Opacity

We consider a process in which a photon of momentum $\mathbf{q}$ and helicity λ is absorbed by a non-relativistic free electron of momentum $\mathbf{p}$ in the neighborhood of an atomic nucleus, giving the electron a non-relativistic momentum $\mathbf{p}'$. The nucleus serves to provide a potential $V(\mathbf{x})$, but is supposed to be so heavy that it can carry away momentum without receiving appreciable energy, so that $p'^2/2m_e = p^2/2m_e + qc$ (where $q \equiv |\mathbf{q}|$, $p \equiv |\mathbf{p}|$, and $p' \equiv |\mathbf{p}'|$) but $\mathbf{p}' \neq \mathbf{p} + \mathbf{q}$. For the present we will consider a general potential, but will later

1.4 Opacity

specialize to a screened Coulomb potential with $V(\mathbf{x}) = -Ze^2 \exp(-r/\ell)/r$ where $r \equiv |\mathbf{x}|$, including the unscreened case where the screening radius ℓ is taken to be infinite.

According to the general rules of quantum mechanics,[17] the differential rate for this process is given by

$$d\Gamma(\mathbf{p} + (\mathbf{q}, \lambda) \to \mathbf{p}') = (2\pi\hbar)^5 n_N \left| M(\mathbf{p} + (\mathbf{q}, \lambda) \to \mathbf{p}') \right|^2$$
$$\times \delta(p'^2/2m_e - p^2/2m_e - qc) d^3p', \quad (1.4.\text{A}1)$$

and so the rate of photon absorption is

$$\Gamma_{\text{abs}}(\mathbf{q}, \lambda) = (2\pi\hbar)^5 n_N \int n_e(\mathbf{p}) d^3p$$
$$\times \int d^3p' \left| M(\mathbf{p} + (\mathbf{q}, \lambda) \to \mathbf{p}') \right|^2 \delta(p'^2/2m_e - p^2/2m_e - qc), \quad (1.4.\text{A}2)$$

where $n_e(\mathbf{p}) d^3p$ is the number density of initial electrons with momenta in a range d^3p around $\mathbf{p}$; n_N is the number density of nuclei; M is the coefficient of the energy and momentum conservation delta functions in the S-matrix element for this process; and we have used the momentum conservation delta function in the rate to eliminate the integral over the final nucleus momentum.

We are only concerned with single-photon absorption processes, and will ignore all quantum electrodynamic radiative corrections, so the matrix element M is of first order in the interaction between the electron and the quantized electromagnetic field. It therefore takes the form[18]

$$M = \frac{-2\pi i}{\sqrt{2qc}(2\pi\hbar)^{3/2}} \times \frac{-\sqrt{4\pi}e\hbar^2}{m_e} \int d^3x \; \psi'^*(\mathbf{x}) \mathbf{e}(\hat{q}, \lambda) \cdot \nabla \psi(\mathbf{x}). \quad (1.4.\text{A}3)$$

Here ψ and ψ' are "in" and "out" solutions of the Schrödinger equations for the initial and final electrons

$$-\frac{\hbar^2}{2m_e}\nabla^2\psi + V\psi = \frac{p^2}{2m_e}\psi, \quad -\frac{\hbar^2}{2m_e}\nabla^2\psi' + V\psi' = \frac{p'^2}{2m_e}\psi', \quad (1.4.\text{A}4)$$

---

[17] For the general relation between S-matrix elements and rates, see e.g. S. Weinberg, *The Quantum Theory of Fields*, Vol. I (Cambridge University Press, Cambridge, 1995), Section 3.4. Note that in this reference $2\pi M$ was defined as the coefficient of the delta function in the S-matrix, while here this coefficient is just M.

[18] For a textbook derivation of this interaction, see e.g. S. Weinberg, *Lectures on Quantum Mechanics*, 2nd edn. (Cambridge University Press, Cambridge, 2015), Eq. 11.7.6. In Eq. (1.4.A3) we are using the electric dipole approximation, in which the photon wavelength is much larger than the de Broglie wavelengths of the initial and final electrons. With photon and electron energies of order $k_B T$, this is a good approximation if $k_B T \ll m_e c^2$, as we shall assume is the case.

normalized so that for $r \to \infty$

$$\psi(\mathbf{x}) \to \frac{\exp(i\mathbf{p}\cdot\mathbf{x}/\hbar)}{(2\pi\hbar)^{3/2}} + O(1/r), \quad \psi'(\mathbf{x}) \to \frac{\exp(i\mathbf{p}'\cdot\mathbf{x}/\hbar)}{(2\pi\hbar)^{3/2}} + O(1/r),$$
(1.4.A5)

where the $O(1/r)$ term is an outgoing wave for ψ and an incoming wave for ψ'. (For an unscreened Coulomb potential the arguments of the exponentials contain additional imaginary terms of order $\ln r$.) Also $e(\hat{q}, \lambda)$ is the polarization vector for a photon with direction $\hat{q}$ and helicity λ, normalized so that $\mathbf{e}^* \cdot \mathbf{e} = 1$. We will use the results for M obtained here also in the discussions of bremsstrahlung in Sections 3.3 and 3.7.

Eventually we will be moving on to the Born approximation, in which M is calculated only to first order in V, but it is useful for several reasons to work for a while with Eq. (1.4.A3), which is derived in what is called the *distorted wave Born approximation*;[19] it is valid to all orders in V but only to first order in the interaction of the electron with the annihilation part of the quantized electromagnetic field.

Multiplying Eq. (1.4.A3) with $qc = p'^2/2m_e - p^2/2m_e$ and using the Schrödinger equations (1.4.A4), we have

$$qcM = \frac{-2\pi i}{\sqrt{2qc}(2\pi\hbar)^{3/2}} \times \frac{-\sqrt{4\pi}e\hbar^2}{m_e}$$

$$\times \int d^3x \left[\left(-\frac{\hbar^2}{2m_e}\nabla^2\psi' + V\psi' \right)^* \mathbf{e}(\hat{q},\lambda)\cdot\nabla\psi \right.$$

$$\left. -\psi'^*\mathbf{e}(\hat{q},\lambda)\cdot\nabla\left(-\frac{\hbar^2}{2m_e}\nabla^2\psi + V\psi \right) \right].$$

Integration by parts shows that the kinetic energy terms cancel,[20] while the potential terms cancel except for a term proportional to the gradient of the potential:

$$M = \frac{-ie\sqrt{\hbar}}{(qc)^{3/2}m_e} \int d^3x \; \psi'^*(\mathbf{x})\mathbf{e}(\hat{q},\lambda)\cdot\left[\nabla V(\mathbf{x})\right]\psi(\mathbf{x}). \quad (1.4.A6)$$

We now go over to the Born approximation, keeping only terms of first order in the potential V. Since Eq. (1.4.A6) already has an explicit factor V, in the Born approximation we can ignore V in the wave functions, and use for ψ and ψ' just the plane waves

---

[19] For a general textbook account of this approximation, see Weinberg, *op. cit.* Section 8.6.
[20] The surface term in the integration by parts may be neglected because of its rapid oscillation as $r \to \infty$ when $p' \neq p$.

1.4 Opacity

$$\psi(\mathbf{x}) = \frac{\exp(i\mathbf{p}\cdot\mathbf{x}/\hbar)}{(2\pi\hbar)^{3/2}}, \quad \psi'(\mathbf{x}) = \frac{\exp(i\mathbf{p}'\cdot\mathbf{x}/\hbar)}{(2\pi\hbar)^{3/2}}. \tag{1.4.A7}$$

Equation (1.4.A6) then reads

$$\begin{aligned} M &= \frac{-ie\sqrt{\hbar}}{(2\pi\hbar)^3 (qc)^{3/2} m_e} \int d^3x \, \mathbf{e}(\hat{q},\lambda) \cdot [\nabla V(\mathbf{x})] \exp\left(i(\mathbf{p}-\mathbf{p}')\cdot\mathbf{x}/\hbar\right) \\ &= \frac{-e}{(2\pi\hbar)^3 (qc)^{3/2} m_e \sqrt{\hbar}} \mathbf{e}(\hat{q},\lambda) \cdot (\mathbf{p}-\mathbf{p}') \\ &\quad \times \int d^3x \, V(\mathbf{x}) \exp\left(i(\mathbf{p}-\mathbf{p}')\cdot\mathbf{x}/\hbar\right). \end{aligned} \tag{1.4.A8}$$

For the screened Coulomb potential $V(\mathbf{x}) = -Ze^2 e^{-r/\ell}/r$, this reads

$$M = \frac{Ze^3}{(2\pi\hbar)^3 (qc)^{3/2} m_e \sqrt{\hbar}} \frac{4\pi \mathbf{e}(\hat{q},\lambda)\cdot(\mathbf{p}-\mathbf{p}')}{(\mathbf{p}-\mathbf{p}')^2/\hbar^2 + 1/\ell^2}. \tag{1.4.A9}$$

Orbital electrons in singly ionized atoms obviously produce a partial screening with ℓ of the order of atomic dimensions. But even where ionization is complete, there is a screening due to mobile electrons attracted to the vicinity of the atomic nucleus. This is known as Debye screening, and is discussed in Section 3.7. For the present, we will consider the unscreened case, with ℓ infinite, in which case

$$M = \frac{Ze^3}{2\pi^2 (qc\hbar)^{3/2} m_e} \frac{\mathbf{e}(\hat{q},\lambda)\cdot(\mathbf{p}-\mathbf{p}')}{(\mathbf{p}-\mathbf{p}')^2}. \tag{1.4.A10}$$

The absorption rate per photon is then given by Eqs. (1.4.A2) and (1.4.A10) as

$$\Gamma_{\text{abs}}(\mathbf{q},\lambda) = \int d^3p \, n_e(\mathbf{p}) \int d^2\hat{p}' \, \frac{8\pi e^6 Z^2 \hbar^2 p' n_N}{m_e c^3 q^3} \left[\frac{(\mathbf{p}-\mathbf{p}')\cdot\mathbf{e}(\lambda,\mathbf{q})}{(\mathbf{p}'-\mathbf{p})^2}\right]^2. \tag{1.4.A11}$$

We average over photon helicity and direction, using

$$\frac{1}{2}\sum_{\lambda=\pm 1} \frac{1}{4\pi} \int d^2\hat{q} \, e_i(\lambda,\mathbf{q}) e_j^*(\lambda,\mathbf{q}) = \frac{1}{8\pi} \int d^2\hat{q} \left[\delta_{ij} - q_i q_j/q^2\right] = \frac{1}{3}\delta_{ij}. \tag{1.4.A12}$$

The integral over the direction of the outgoing electron is then

$$\int d^2\hat{p}' \, \frac{1}{(\mathbf{p}-\mathbf{p}')^2} = \frac{2\pi}{pp'} \ln\left(\frac{p'+p}{p'-p}\right). \tag{1.4.A13}$$

Equation (1.4.A11) now gives the average photon absorption rate

$$\Gamma_{\text{abs}}(q) = \int d^3p \, n_e(\mathbf{p}) \frac{16\pi^2 e^6 Z^2 \hbar^2 n_N}{3 p m_e c^3 q^3} \ln\left(\frac{p'+p}{p'-p}\right). \quad (1.4.\text{A}14)$$

This can be rewritten for the purposes of comparison with the main text, setting $v = p/m_e$, $v' = p'/m_e$, $\nu = qc/h$, and $h = 2\pi\hbar$. Equation (1.4.A14) then becomes Eq. (1.4.10).

In this derivation we have treated the Coulomb interaction between electrons and nuclei only to first order in the Coulomb potential. This is justified if Ze^2/r for typical values of r is much less than the electron kinetic energies. Taking the typical value of r as the de Broglie wavelength $\hbar/m_e v$, the ratio of potential to kinetic energy is of order

$$\frac{Ze^2/r}{m_e v^2/2} \approx \frac{Ze^2}{\hbar v} \simeq Zc/137v,$$

so this calculation is reliable only if $v/c \gg Z/137$. Our non-relativistic treatment also requires that $v/c \ll 1$. For nuclei like C, N, and O, with $Z \geq 6$, this does not leave much of a range for the electron velocity in which the above calculation is reliable, beyond just giving the order of magnitude of the absorption rate. The contribution to M of terms of higher order in the Coulomb potential is considered in the context of bremsstrahlung in Section 3.7.

1.5 Nuclear Energy Generation

We now consider the nuclear energy production per mass $\epsilon(\rho, T)$. As with opacity in the previous section, one of our aims here will be to estimate the exponents when $\epsilon(\rho, T)$ is approximated by a power-law expression

$$\epsilon(\rho, T) \simeq \epsilon_1 \rho^\lambda (k_B T)^\nu, \quad (1.5.1)$$

with ϵ_1 as well as λ and ν independent of ρ and T.

The nuclear material left over from the first three minutes of the big bang was chiefly ^{1}H (that is, protons), plus about 25% by mass ^{4}He, and only a trace of ^{2}H, ^{3}He, and ^{7}Li. These light nuclei have less binding energy per nucleon than nuclei of medium atomic weight like iron and nickel, so energy can be gained by fusion of hydrogen and helium into heavier elements. But there are no stable nuclei with five or eight nucleons, so it is difficult (though, as we shall see, not impossible) to gain energy from helium in ^{1}H–^{4}He or ^{4}He–^{4}He collisions. Thus, as long as hydrogen lasts in the center of a star, the dominant source of nuclear energy will be the fusion of ^{1}H into ^{4}He, which has by far the greatest binding energy of any of these light elements.

1.5 Nuclear Energy Generation

There are two chief routes by which hydrogen can fuse into helium. One is the proton–proton chain,[21] of which the simplest version is[22]

$$
\begin{aligned}
&\text{I}: \quad {}^1\text{H} + {}^1\text{H} \to {}^2\text{H} + e^+ + \nu_e + 1.18 \text{ MeV} \\
&\text{II}: \quad {}^1\text{H} + {}^2\text{H} \to {}^3\text{He} + \gamma + 5.49 \text{ MeV} \\
&\text{III}: \quad {}^3\text{He} + {}^3\text{He} \to {}^4\text{He} + {}^1\text{H} + {}^1\text{H} + 12.85 \text{ MeV}.
\end{aligned} \quad (1.5.2)
$$

The other route is the CNO cycle,[23] which in its simplest variant is

$$
\begin{aligned}
&\text{i}: \quad {}^1\text{H} + {}^{12}\text{C} \to {}^{13}\text{N} + \gamma + 1.95 \text{ MeV} \\
&\text{ii}: \quad {}^{13}\text{N} \to {}^{13}\text{C} + e^+ + \nu_e + 1.50 \text{ MeV} \\
&\text{iii}: \quad {}^1\text{H} + {}^{13}\text{C} \to {}^{14}\text{N} + \gamma + 7.54 \text{ MeV} \\
&\text{iv}: \quad {}^1\text{H} + {}^{14}\text{N} \to {}^{15}\text{O} + \gamma + 7.35 \text{ MeV} \\
&\text{v}: \quad {}^{15}\text{O} \to {}^{15}\text{N} + e^+ + \nu_e + 1.73 \text{ MeV} \\
&\text{vi}: \quad {}^1\text{H} + {}^{15}\text{N} \to {}^{12}\text{C} + {}^4\text{He} + 4.96 \text{ MeV},
\end{aligned} \quad (1.5.3)
$$

where carbon, nitrogen, and oxygen nuclei are understood to be present in the interstellar matter from which stars like the Sun are formed, left over from nuclear processes in an earlier generation of stars. They are catalysts, neither created nor destroyed in a complete cycle. In both cases there are side branches and extensions to which we will return below, but these simple versions will provide us with sufficient examples to illustrate how $\epsilon(\rho, T)$ is estimated.

The detailed calculation of the rates of these various nuclear reactions is beyond the scope of this book. However, we can usefully identify various suppression factors in the rates that tell us a good deal about which reactions are dominant, and about their temperature dependence.

Electromagnetic Coupling

The rate of any reaction in which a single photon is emitted (such as ${}^1\text{H} + {}^2\text{H} \to {}^3\text{He} + \gamma$ in the proton–proton cycle or ${}^1\text{H} + {}^{12}\text{C} \to {}^{13}\text{N} + \gamma$ in the CNO cycle) is suppressed by a factor of order $e^2/\hbar c \simeq 1/137$.

Weak Coupling

The rate of any reaction in which a proton turns into a neutron with the emission of a positron and neutrino (such as the first step ${}^1\text{H} + {}^1\text{H} \to {}^2\text{H} + e^+ + \nu_e$ in

---

[21] H. A. Bethe and C. H. Critchfield, *Phys. Rev.* **54**, 248 (1938).
[22] The energies listed here for the proton–proton chain and below for the CNO cycle are the energies for each reaction actually deposited in the stellar material. Thus, where a positron is emitted, these energies include not only the rest energy $m_e c^2$ of the emitted positron but also the rest energy of the electron with which that positron inevitably annihilates. On the other hand, the mean energy of the accompanying neutrino is subtracted from the energy released, since virtually all neutrinos leave the star.
[23] C. F. von Weizsäcker, *Phys. Z.* **38**, 176 (1938); H. A. Bethe, *Phys. Rev.* **55**, 434 (1939).

the proton–proton cycle or the beta decays of ^{13}N and ^{15}O in the CNO cycle) is suppressed by two factors of the weak coupling constant $G_{wk} = 1.1664 \times 10^{-11}$ MeV^{-2}. Since the typical energy involved in these nuclear reactions is about 1 MeV, weak interaction processes are typically suppressed by a dimensionless factor of order 10^{-22}.

Coulomb Barrier

The temperature dependence of nuclear reaction rates is chiefly due to the necessity for colliding nuclei to leak through the Coulomb barrier, the field of electrostatic repulsion between positively charged atomic nuclei.[24] The calculation of the effect of the Coulomb barrier on reaction rates requires use of quantum mechanics, but only at a quite elementary level, and will be presented in an appendix at the end of this section. The result is that a reaction involving two nuclei of atomic numbers Z_1 and Z_2 and an energy of relative motion E is suppressed by a factor of order

$$B(E) = \exp\left[-\pi Z_1 Z_2 e^2 \sqrt{\frac{2\mu}{\hbar^2 E}}\right], \qquad (1.5.4)$$

where $\mu = m_1 m_2/(m_1 + m_2)$ is the reduced mass.

The nuclei colliding in a star of course do not have any definite value for the energy E of relative motion, but rather a range of values, with probabilities governed by the requirements of kinetic theory at temperature T. Assuming that nuclei spend most of their time sufficiently far from other nuclei that their energy is mostly kinetic, the probability of finding a pair of nuclei in a range of momenta $d^3 p_1 \, d^3 p_2$ is proportional to

$$\exp\left(-\frac{\mathbf{p_1}^2}{2m_1 k_B T} - \frac{\mathbf{p_2}^2}{2m_2 k_B T}\right) d^3 p_1 \, d^3 p_2 = \exp\left(-\frac{E}{k_B T}\right) d^3 p$$

$$\times \exp\left(-\frac{\mathbf{P}^2}{2(m_1 + m_2)k_B T}\right) d^3 P, \qquad (1.5.5)$$

where $\mathbf{P} \equiv \mathbf{p}_1 + \mathbf{p}_2$ is the total momentum, and $E = \mathbf{p}^2/2\mu$ is the energy of relative motion, with $\mathbf{p} \equiv \mu(\mathbf{p}_1/m_1 - \mathbf{p}_2/m_2)$ the relative momentum. The rate ϵ of nuclear reactions per gram is then of the form

$$\epsilon(\rho, T) = \int_0^\infty dE \, f(E, \rho, T) \exp(-E/k_B T) B(E)$$

$$= \int_0^\infty dE \, f(E, \rho, T) \exp\left(-\frac{E}{k_B T} - \frac{C}{\sqrt{E}}\right), \qquad (1.5.6)$$

---

[24] Barrier penetration was first calculated in the context of nuclear α-decay; G. Gamow, *Z. Phys.* **52**, 510 (1928).

1.5 Nuclear Energy Generation

where $f(E, \rho, T)$ arises from power-law factors in the thermal distribution of E and $\mathbf{P}$ and in the probability of the nuclear reaction occurring when the nuclei reach zero separation, and C is the constant in the exponent in Eq. (1.5.4):

$$C = \pi Z_1 Z_2 e^2 \sqrt{\frac{2\mu}{\hbar^2}}. \tag{1.5.7}$$

In practice, $k_B T$ is always much less than C^2, so the exponential $\exp(-C/\sqrt{E})$ will be very small unless E is much greater than $k_B T$, in which case $\exp(-E/k_B T)$ will be very small. The exponential in Eq. (1.5.6) is therefore very sharply peaked at the energy E_T where its argument is a maximum:

$$0 = \frac{d}{dE}\bigg|_{E=E_T} \left(-\frac{E}{k_B T} - \frac{C}{\sqrt{E}}\right) = -\frac{1}{k_B T} + \frac{C}{2E_T^{3/2}} \tag{1.5.8}$$

so

$$E_T = (C k_B T / 2)^{2/3}. \tag{1.5.9}$$

The dominant factor B_T in the temperature dependence of the reaction rate (1.5.6) is simply the exponential function, evaluated at $E = E_T$:

$$B_T = \exp\left(-\frac{E_T}{k_B T} - \frac{C}{\sqrt{E_T}}\right) = \exp\left(-3\left(\frac{\pi Z_1 Z_2 e^2 \sqrt{\mu}}{\hbar\sqrt{2k_B T}}\right)^{2/3}\right). \tag{1.5.10}$$

Numerically this is

$$B_T = \exp\left[-\left(Z_1^2 Z_2^2 (\mu/m_p) \times \frac{7.726 \times 10^{10}\,\text{K}}{T}\right)^{1/3}\right], \tag{1.5.11}$$

where m_p is the proton mass.

The values of reaction rates depend on a number of other factors besides the barrier penetration factor, but it is the barrier that chiefly governs their temperature dependence. Thus we can use the above calculation of the Coulomb barrier to estimate the exponent ν in the power law $\epsilon \propto (k_B T)^\nu$ that is used to estimate the temperature dependence of the energy generation rate ϵ. We take

$$\nu = T \frac{d}{dT} \ln B_T \simeq \frac{1}{3}\left(Z_1^2 Z_2^2 (\mu/m_p) \times \frac{7.726 \times 10^{10}\,\text{K}}{T}\right)^{1/3}. \tag{1.5.12}$$

(The $T^{-1/3}$ temperature dependence here is sufficiently weak to justify approximating ϵ as proportional to a constant power of temperature.) From Eqs. (1.5.11) and (1.5.12) we infer the general rule that ν is one-third the absolute value of the exponent in whatever barrier penetration factor dominates the temperature dependence of the energy generation rate.

Let us now apply these general remarks to stars that derive their nuclear energy either from the proton–proton chain or from the CNO cycle.

Proton–Proton Chain

For the first reaction $p + p \to d + e^+ + \nu$ in the proton–proton chain we take $\mu = m_p/2$ and $Z_1 = Z_2 = 1$, so, according to Eq. (1.5.11), if $T = 10^7$ K (roughly the temperature at the center of the Sun), the Coulomb barrier suppresses the reaction by a factor $\exp(-15.7) = 1.5 \times 10^{-7}$.

But the reaction $p + p \to d + e^+ + \nu$ is not the end of the story; it is just the first step in a chain of reactions. The Coulomb barrier suppression of the second step, $^1\text{H} + {}^2\text{H} \to {}^3\text{He} + \gamma$, is only slightly more severe than that of reaction I, because the charges of the nuclei are the same, and their reduced mass is larger only by a factor $4/3$. Taking $\mu = 2m_p/3$, $Z_1 = Z_2 = 1$, and $T \simeq 10^7$ K in Eq. (1.5.11) gives $B_T \approx \exp(-4/3 \times 15.7) = 8 \times 10^{-10}$. Apart from Coulomb suppression, since step I involves a weak interaction it is suppressed by an additional factor of order 10^{-22} and since step II involves an electromagnetic interaction it is suppressed by an additional factor of order $1/137$, so the ratio of the rate per proton of step I and the rate per deuteron of step II is expected to be of order

$$\frac{\text{rate/p of } p + p \to d + e^+ + \nu}{\text{rate/d of } p + d \to {}^3\text{He} + \gamma} \approx \frac{10^{-22} \times (1.5 \times 10^{-7})}{(1/137) \times (8 \times 10^{-10})} \simeq 3 \times 10^{-18}.$$

(The actual ratio is about 10^{-17}.) Reaction III has a more formidable Coulomb barrier, with $Z_1 Z_2 = 4$. All three reactions release substantial amounts of energy. So which do we need to calculate in order to find ϵ. And in particular, which is the relevant Coulomb barrier?

The answer relies on an assumption of time-independence: The abundances of the intermediate participants in these reactions rapidly evolve to stable values, for which these abundances change little over times in which a very large number of reactions take place in the star's core. Thus, in order that the abundance of deuterons should not change, the rates per volume of reactions I and II, in which deuterons are respectively created and destroyed, should be the same, and in order that the abundance of ^{3}He nuclei should not change, the rate per volume of reaction II should be twice that of reaction III, in which two ^{3}He nuclei are destroyed:

$$\Gamma \equiv \Gamma(\text{I}) = \Gamma(\text{II}) = 2\Gamma(\text{III}), \qquad (1.5.13)$$

where the Γs denote the rates per volume of various reactions. It is like the law of economics that supply equals demand. If demand exceeds supply prices will go up, damping demand and providing an incentive for increased supply, until supply and demand approach each other. (Or so they say.) In the same way, if the rate per volume of reaction II were less than that of reaction I the abundance of ^{2}H nuclei would rise until these rates were equal, and just as many ^{2}H nuclei were being destroyed as created. According to the above estimate of the ratio of the rate per proton of reaction I and the rate per deuteron of reaction II, we therefore expect the number density of deuterons to be smaller than the number density of protons by a factor of order 3×10^{-18}.

Though $\Gamma(\text{I})$, $\Gamma(\text{II})$, and $2\Gamma(\text{III})$ must all be equal, their calculation differs in one important respect. The rate of reaction I does not depend on the abundance of the intermediate nuclei ^2H and ^3He, and in particular is not suppressed by their low abundance, so it can be calculated without knowing anything about the other reactions. Thus it is the Coulomb barrier in reaction I that governs the rate at which hydrogen is converted to helium and energy is produced, and its temperature dependence. In particular, in accordance with the general rule (1.5.12), for the proton–proton cycle the exponent ν in the temperature dependence of ϵ is one-third of the value 15.7 that we previously calculated for the exponent in the barrier penetration factor for reaction I, so $\nu \simeq 5$ at $T \approx 10^7$ K. Fortunately ν has only a mild dependence on temperature, going as $T^{-1/3}$, so this estimate of ν is a fair approximation for a wide range of temperatures.

But although we only need to calculate the rate Γ of reaction I, all of reactions I, II, and III release energy, say an energy E_I, E_II, and E_III per reaction, so the rate $\epsilon\rho$ of total energy production per volume is not just $E_\text{I}\Gamma$, but

$$\epsilon\rho = \left(E_\text{I} + E_\text{II} + \frac{1}{2}E_\text{III}\right)\Gamma = 13.1 \text{ MeV} \times \Gamma. \quad (1.5.14)$$

The crucial first step in the proton–proton chain is a collision of two protons. Its rate, and hence the rate per volume $\epsilon\rho$ of energy generation due to the proton–proton chain, is proportional to ρ^2. Hence, if the proton–proton chain dominates nuclear energy generation, we have $\lambda = 1$ as well as $\nu \approx 5$.

The reactions (1.5.2) dominate the energy production in the proton–proton chain, but there are alternative finales to this chain, one of which is of historical importance. In one alternative, instead of a pair of ^3He nuclei combining in reaction III, individual ^3He nuclei undergo the reaction

$$\text{III}': \quad {}^3\text{He} + {}^4\text{He} \to {}^7\text{Be} + \gamma$$

followed by either

$$\begin{aligned}\text{IV}: & \quad {}^7\text{Be} + e^- \to {}^7\text{Li} + \nu_e \\ \text{V}: & \quad {}^7\text{Li} + {}^1\text{H} \to {}^4\text{He} + {}^4\text{He}\end{aligned} \quad (1.5.15)$$

or else

$$\begin{aligned}\text{IV}': & \quad {}^7\text{Be} + {}^1\text{H} \to {}^8\text{B} + \gamma \\ \text{V}': & \quad {}^8\text{B} \to {}^8\text{Be} + e^+ + \nu_e \\ \text{VI}': & \quad {}^8\text{Be} \to {}^4\text{He} + {}^4\text{He}.\end{aligned} \quad (1.5.16)$$

The probability of a ^3He nucleus undergoing the reaction III' rather than III is small, so these alternatives have little effect on the energy generation rate ϵ

and its density and temperature dependence, but the high energy of the neutrino from the ^{8}B beta decay in reaction V′, extending up to over 10 MeV, offered an early opportunity of observing neutrinos from the Sun.

The reaction $^{37}\text{Cl} + \nu_e \rightarrow {}^{37}\text{Ar} + e^-$ that was used to search for solar neutrinos in the experiments of Davis *et al.*[25] on solar neutrinos is sensitive only to these high-energy neutrinos, not to the much lower-energy neutrinos emitted in the other reactions of the proton–proton chain. The high Coulomb barriers in reactions III′ and IV′ make the flux of high-energy neutrinos extremely sensitive to the temperature profile in the Sun. Detailed calculations by John Bahcall[26] (1934–2005) showed that the high-energy neutrinos should be observable in Davis's experiments, but decades of searching did not find them. Finally solar neutrinos were detected[27] using the reaction $\text{Ga}^{71} + \nu_e \rightarrow \text{Ge}^{71} + e^-$, but the observed rate was substantially less than predicted by Bahcall. Either Bahcall's calculations were inaccurate, or something was happening to neutrinos on the way to the Earth.

In particular, it was speculated by Bruno Pontecorvo (1913–1993) that neutrinos have mass, and that the states of definite mass are not the neutrinos of electron type emitted in the Sun, but superpositions of neutrinos of electron type with neutrinos of muon and tauon type, so that on the way to Earth electron-type neutrinos become an oscillating superposition of types, with only the electron-type fraction observable in reactions like $\text{Cl}^{37} + \nu_e \rightarrow \text{Ar}^{37} + e^-$ or $\text{Ga}^{71} + \nu_e \rightarrow \text{Ge}^{71} + e^-$. The issue was settled by experiments at the Sudbury Neutrino Observatory.[28] By monitoring a large tank of heavy water, experimenters could detect high-energy ^{8}B neutrinos not only in the reaction $\nu_e + d \rightarrow p + p + e^-$, which is sensitive only to electron-type neutrinos, but also in the neutral current process $\nu + d \rightarrow p + n + \nu$, which is equally sensitive to neutrinos of all types, electron, muon, and tauon. It turned out that the total flux of neutrinos of all types agreed with Bahcall's calculations, providing a decisive vote in favor of neutrino oscillations. Since then the existence of neutrino oscillations has been confirmed and neutrino masses and mixing angles measured in numerous terrestrial experiments.

CNO Cycle

Matters are more complicated for the CNO cycle. Here too we assume that the abundances of the intermediate CNO nuclei settle down to constant values. The constancy of the abundance of ^{13}N requires that reactions i and ii have the same rate per volume; the constancy of the abundance of ^{13}C requires that reactions

---

[25] R. Davis, D. S. Harmer, and K. C. Hoffman, *Phys. Rev. Lett.* **20**, 1205 (1968).
[26] J. N. Bahcall, *Current Science* **77**, 1487 (1999), and earlier references quoted therein.
[27] P. Anselmann *et al.*, *Phys. Lett.* **B342**, 440 (1995); J. N. Abdurashitov *et al.*, *Phys. Rev. Lett.* **77**, 3708 (1996).
[28] Q. R. Ahmad *et al.*, *Phys. Rev. Lett.* **89**, 11301 (2002).

1.5 Nuclear Energy Generation

ii and iii have the same rate per volume; and so on, so that all these rates per volume are equal:

$$\Gamma(\text{i}) = \Gamma(\text{ii}) = \Gamma(\text{iii}) = \Gamma(\text{iv}) = \Gamma(\text{v}) = \Gamma(\text{vi}) \equiv \Gamma. \qquad (1.5.17)$$

This determines the *ratios* of the abundances. Each of the rates here is proportional to the number density n of the CNO nucleus in the initial state of the reaction

$$\Gamma(\text{i}) = n(^{12}\text{C})R(\text{i}), \quad \Gamma(\text{ii}) = n(^{13}\text{N})R(\text{ii}), \text{ etc.,} \qquad (1.5.18)$$

with the rate factors R independent of the densities of anything but hydrogen. For each reaction, R is the rate at which the CNO nucleus in the initial state undergoes that reaction. For instance, $R(\text{i})$ is the rate at which any individual ^{12}C nucleus undergoes the reaction $^{1}\text{H} + ^{12}\text{C} \rightarrow ^{13}\text{N} + \gamma$. Then the equality of rates (1.5.17) gives

$$\frac{n(^{13}\text{N})}{n(^{12}\text{C})} = \frac{R(\text{i})}{R(\text{ii})}, \quad \frac{n(^{13}\text{C})}{n(^{12}\text{C})} = \frac{R(\text{i})}{R(\text{iii})}, \quad \text{etc.} \qquad (1.5.19)$$

But we cannot in this way find the overall number density of the CNO nuclei

$$n(\text{CNO}) \equiv n(^{12}\text{C}) + n(^{13}\text{N}) + n(^{13}\text{C}) + n(^{14}\text{N}) + n(^{15}\text{O}) + n(^{15}\text{N}), \quad (1.5.20)$$

which does not change in the reactions i through vi, and is determined by the abundances in the interstellar medium from which the star formed. We can, however, express the common rate Γ in terms of $n(\text{CNO})$: Using Eqs. (1.5.20) and (1.5.17) and then (1.5.18) we note that

$$\frac{n(\text{CNO})}{\Gamma} = \frac{n(^{12}\text{C})}{\Gamma(\text{i})} + \frac{n(^{13}\text{N})}{\Gamma(\text{ii})} + \frac{n(^{13}\text{C})}{\Gamma(\text{iii})} + \frac{n(^{14}\text{N})}{\Gamma(\text{iv})} + \frac{n(^{15}\text{O})}{\Gamma(\text{v})} + \frac{n(^{15}\text{N})}{\Gamma(\text{vi})}$$

$$= \frac{1}{R(\text{i})} + \frac{1}{R(\text{ii})} + \frac{1}{R(\text{iii})} + \frac{1}{R(\text{iv})} + \frac{1}{R(\text{v})} + \frac{1}{R(\text{vi})},$$

so the common rate is

$$\Gamma = n(\text{CNO}) \bigg/ \left(\frac{1}{R(\text{i})} + \frac{1}{R(\text{ii})} + \frac{1}{R(\text{iii})} + \frac{1}{R(\text{iv})} + \frac{1}{R(\text{v})} + \frac{1}{R(\text{vi})} \right). \qquad (1.5.21)$$

That is, the common rate of the reactions equals the harmonic mean of what the individual rates would be if the density of the CNO nucleus in each initial state equaled the total density $n(\text{CNO})$. The rate per volume $\epsilon\rho$ of energy generation in the CNO cycle is Γ times the sum of the energies in Eq. (1.5.3):

$$\epsilon\rho = \Gamma \times 25.03 \text{ MeV.} \qquad (1.5.22)$$

Because of the absence of a Coulomb barrier in the beta decays ii and v, these reactions have relatively rapid rates R per CNO nucleus, with mean lives $1/R$

of 7 minutes and 82 seconds, respectively, while $1/R$ for all the other reactions in the CNO cycle is at least 10^5 years. Thus the terms $1/R(\text{ii})$ and $1/R(\text{v})$ can be neglected in the denominator in Eq. (1.5.21). Also, for the same reason, the number density of the CNO nucleus in the initial states of the beta decay reactions is much smaller than the number densities of the other CNO nuclei, and can be neglected in $n(\text{CNO})$. Thus Eq. (1.5.21) for the rate Γ of the various reactions in the CNO channel is dominated by the two-body reactions i, iii, iv, and vi. As two-body reactions, they all have $\lambda = 1$. Also, these reactions all have about the same value of the reduced mass, ranging from $12 m_\text{p}/13$ to $15 m_\text{p}/16$, while $Z_1 Z_2$ only ranges from 6 for reaction i to 7 for reaction vi, so the Coulomb suppression factor and hence the rate factor R is smallest for reaction vi, but not overwhelmingly so. We will take the Coulomb barriers of these reactions to be a compromise, calculated by taking $Z_1 Z_2 = 6.5$ and $\mu = m_\text{p}$. At any given temperature, the exponent in Eq. (1.5.10) for the effective Coulomb barrier is thus larger than for the proton–proton chain by a factor $6.5^{2/3} 2^{1/3} = 4.4$. At the nominal temperature of 10^7 K, the Coulomb barrier in the CNO cycle produces a suppression factor $\exp(-4.4 \times 15.7) \simeq 10^{-30}$. It is only because of the extreme slowness of weak interaction processes such as the first step in the proton–proton chain that the CNO cycle can compete with the proton–proton chain at any temperature.

The power of temperature in Eq. (1.5.1) is larger than for the proton–proton chain by the same factor 4.4, so at $T \approx 10^7$ K we have $\nu \approx 22$, and somewhat less at higher temperatures. As already mentioned, the power of density is $\lambda = 1$.

Here too there are alternative finales. Instead of step vi, the ^{15}N nucleus can undergo the reaction $^1\text{H} + {}^{15}\text{N} \to {}^{16}\text{O} + \gamma$, followed by $^1\text{H} + {}^{16}\text{O} \to {}^{17}\text{F} + \gamma$ and $^{17}\text{F} \to {}^{17}\text{O} + e^+ + \nu$. After that, there are again two possibilities: either $^1\text{H} + {}^{17}\text{O} \to {}^{14}\text{N} + {}^4\text{He}$, or else $^1\text{H} + {}^{17}\text{O} \to {}^{18}\text{F} + \gamma$ followed by $^{18}\text{F} \to {}^{18}\text{O} + e^+ + \nu$ and $^1\text{H} + {}^{18}\text{O} \to {}^{15}\text{N} + {}^4\text{He}$. In all cases the net effect is that four protons turn into a ^4He nucleus plus two positrons and two neutrinos, with the CNO catalysts always returned to their original abundances.

Crossover

We can now estimate the crossover temperature at which the rates of energy production in the CNO cycle and proton–proton chain would be equal. We have seen that the rate of the reactions in the proton–proton chain is suppressed by the Coulomb barrier by a factor $\exp\left(-15.7 (T\,[10^7\,\text{K}])^{-1/3}\right)$, so the rate of the reactions in the CNO cycle is suppressed by a factor $\exp\left(-4.4 \times 15.7 (T\,[10^7\,\text{K}])^{-1/3}\right)$. It is further suppressed relative to the proton–proton chain by the ratio of the number of CNO nuclei to hydrogen nuclei, which for the Sun is about 10^{-3}, and since a photon is emitted, also by a factor $e^2/\hbar c \simeq 10^{-2}$. On the other hand, the reaction $p + p \to d + e^+ + \nu$ in the proton–proton chain is a weak interaction, so its rate is proportional to the square of the weak coupling constant, and is therefore suppressed by a dimensionless factor $(G_{\rm wk}E^2)^2$, which for $E \approx 1$ MeV is about 10^{-22}. So, very roughly, the crossover temperature at which the CNO cycle and the proton–proton chain have competitive rates is given by

$$10^{-3} \times 10^{-2} \times \exp\left(-4.4 \times 15.7(T\,[10^7\,{\rm K}])^{-1/3}\right)$$
$$\approx 10^{-22} \times \exp\left(-15.7(T\,[10^7\,{\rm K}])^{-1/3}\right),$$

or $T \approx 2.5 \times 10^7$ K. This is not very different from the value given by more detailed calculations,[29] which is not much greater than the temperature 1.36×10^7 K at the center of the Sun. For stars that are considerably more or less massive than the Sun the central temperature is higher or lower, and it is respectively the CNO cycle or the proton–proton chain that dominates energy production.

Beyond Hydrogen Burning

As mentioned in Section 1.3, when the hydrogen has been converted to helium in a star's center, the star leaves the main sequence and becomes a red giant, in which the conversion of hydrogen to helium continues in a shell surrounding the helium core. The core temperature continues to grow, and when it becomes sufficiently high it becomes the turn of helium to undergo nuclear reactions. Although there is no stable nucleus that can be formed in a collision of a proton and a ^{4}He nucleus or in the collision of two ^{4}He nuclei, the latter collision can produce an unstable state of the nucleus ^{8}Be that lives long enough before it undergoes fission back into two ^{4}He nuclei, so that it can serve as an intermediary in the carbon production reactions

$$\begin{aligned} {\rm a}: \quad & ^4{\rm He} + {}^4{\rm He} \to {}^8{\rm Be} + \gamma \\ {\rm b}: \quad & ^4{\rm He} + {}^8{\rm Be} \to {}^{12}{\rm C} + \gamma. \end{aligned} \quad (1.5.23)$$

Although this is a sequence of two-body reactions, it does not lead to an energy production rate per volume $\epsilon\rho$ proportional to ρ^2, as in the proton–proton chain and the CNO cycle. The reason is that there is only a small probability $\mathcal{P}$ for the ^{8}Be nucleus to absorb another ^{4}He nucleus before it fissions. Thus $\epsilon\rho$ is proportional to $\rho^2\mathcal{P}$, and since $\mathcal{P}$ when small is proportional to ρ, $\epsilon\rho$ is proportional to ρ^3, and therefore the exponent λ in Eq. (1.5.1) is $\lambda = 2$.

As usual, the temperature dependence of ϵ is harder to estimate. Reaction a is endothermic, requiring an energy E of relative motion of the two ^{4}He nuclei of at least 92 keV. In order for ^{4}He nuclei to have any chance of having energies this large, the temperature must be at least 10^8 K. Even at such relatively high

---

[29] R. J. Tayler, *The Stars: Their Structure and Evolution* (Wykeham Publications, London, 1970), Figure 39m gives the crossover temperature as 1.7×10^7 K, while F. LeBlanc, *An Introduction to Stellar Astrophysics* (John Wiley & Sons, Winchester, 2010), Figure 6.7 gives 1.9×10^7 K.

temperatures, there are sizable Coulomb barriers both in the rate for reaction a and in the probability $\mathcal{P}$ that a ^{8}Be nucleus will experience reaction b instead of fissioning. The only reason[30] why carbon production is non-negligible at temperatures of order 10^8 K to 10^9 K is that there is an unstable state of ^{12}C that provides a resonance in the ^{4}He+^{8}Be channel at an accessible excitation energy of 310 keV. This unstable state has an appreciable chance of decaying into the stable ground state of carbon, with the emission of a 7.4 MeV photon. Because of the pair of Coulomb barriers plus the exothermic nature of reaction a, the exponent ν in Eq. (1.5.1) for the temperature dependence of carbon production is quite large, estimated to be of order 30 to 40, depending on the temperature.

Once ^{12}C is formed in this way, it is possible to produce heavier nuclei in various reactions that are suppressed mostly by Coulomb barriers: ^{4}He+^{12}C $\rightarrow$ ^{16}O+γ, ^{4}He+^{16}O $\rightarrow$ ^{24}Mg+γ, ^{12}C+^{12}C $\rightarrow$ ^{24}Mg+γ, and so on. There are also reactions that destroy but do not produce various light nuclei with relatively small binding energies, including ^{2}H, ^{3}He, ^{6}Li, ^{7}Li, ^{9}Be, ^{10}B, and ^{11}B. Where these nuclei are found spectroscopically in interstellar clouds, their measured abundance provides a valuable lower bound on the cosmological abundance of light elements left over from the beginning of the big bang.

Appendix: Calculation of Suppression by Coulomb Barriers

Classically, the total energy of a pair of nuclei interacting through a central potential $V(r)$ is

$$E_{\text{tot}} = \frac{\mathbf{p}_1^2}{2m_1} + \frac{\mathbf{p}_2^2}{2m_2} + V(r) = \frac{\mathbf{P}^2}{2(m_1+m_2)} + \frac{\mathbf{p}^2}{2\mu} + V(r), \qquad (1.5.\text{A1})$$

where $\mathbf{p}$ and $\mathbf{P}$ are the relative and total momenta, where

$$\mathbf{p} = \mu\left(\frac{\mathbf{p}_1}{m_1} - \frac{\mathbf{p}_2}{m_2}\right), \quad \mathbf{P} = \mathbf{p}_1 + \mathbf{p}_2, \qquad (1.5.\text{A2})$$

and μ again is the reduced mass

$$\mu = \frac{m_1 m_2}{m_1 + m_2}.$$

Since both E_{tot} and $\mathbf{P}$ are time-independent, they can be expressed at any time in terms of the relative and total momenta $\mathbf{p}_0$ and $\mathbf{P}_0$ at a time t_0 early enough that the nuclei are so far apart that $V(r)$ is negligible:

$$E_{\text{tot}} = \frac{\mathbf{P}_0^2}{2(m_1+m_2)} + \frac{\mathbf{p}_0^2}{2\mu}, \quad \mathbf{P} = \mathbf{P}_0. \qquad (1.5.\text{A3})$$

---

[30] E. E. Salpeter, *Astrophys. J.* **115**, 326 (1952).

1.5 Nuclear Energy Generation

We are assuming here that the potential depends only on the separation $r \equiv |\mathbf{x}|$, $\mathbf{x} = \mathbf{x}_1 - \mathbf{x}_2$.

Quantum mechanically, the probability of finding the nuclei with separation vector $\mathbf{x}_1 - \mathbf{x}_2$ in a small volume $d^3\mathbf{x}$ around $\mathbf{x}$ and center-of-mass position $(m_1\mathbf{x}_1 + m_2\mathbf{x}_2)/(m_1 + m_2)$ in a small volume $d^3\mathbf{X}$ around $\mathbf{X}$ is given in terms of a wave function $\psi(\mathbf{x}, \mathbf{X})$ by $|\psi(\mathbf{x}, \mathbf{X})|^2 d^3\mathbf{x}\, d^3\mathbf{X}$. The wave function satisfies the Schrödinger equation $H\psi = E_{\text{tot}}\psi$, where E_{tot} is the numerical quantity given by Eq. (1.5.A3), and H is the Hamiltonian operator, given by replacing $\mathbf{p}$ and $\mathbf{P}$ on the right-hand side of Eq. (1.5.A1) with $-i\hbar$ times gradients with respect to the separation $\mathbf{x}$ and the center-of-mass position $\mathbf{X} = (m_1\mathbf{x}_1 + m_2\mathbf{x}_2)/(m_1 + m_2)$. The Schrödinger equation is then

$$E_{\text{tot}}\psi(\mathbf{x}, \mathbf{X}) = \left[-\frac{\hbar^2}{2(m_1 + m_2)}\nabla_{\mathbf{X}}^2 - \frac{\hbar^2}{2\mu}\nabla_{\mathbf{x}}^2 + V(r)\right]\psi(\mathbf{x}, \mathbf{X}). \quad (1.5.\text{A4})$$

We can always find a solution of the form

$$\psi(\mathbf{x}, \mathbf{X}) = e^{i\mathbf{P}\cdot\mathbf{X}/\hbar}\psi_E(\mathbf{x}), \quad (1.5.\text{A5})$$

where E is the energy of relative motion, defined by

$$E_{\text{tot}} = \frac{\mathbf{P}^2}{2(m_1 + m_2)} + E, \quad (1.5.\text{A6})$$

and

$$E\psi_E(\mathbf{x}) = \left[-\frac{\hbar^2}{2\mu}\nabla_{\mathbf{x}}^2 + V(r)\right]\psi_E(\mathbf{x}). \quad (1.5.\text{A7})$$

This is supposed to hold only outside some very small radius r_0, within which nuclear reactions occur.

To solve this equation, we can often employ the WKB approximation. We suppose that for a range of radii $r > r_0$, $V(r) - E$ is positive and sufficiently large that $V(r)$ changes little in a distance $1/\kappa_E(r)$, where

$$\kappa_E(r) = \left[\frac{2\mu}{\hbar^2}(V(r) - E)\right]^{1/2}.$$

Then, in this range of r,

$$\psi_E(r) \simeq C_+ \exp\left(+\int_{r_0}^r \kappa_E(r')\,dr'\right) + C_- \exp\left(-\int_{r_0}^r \kappa_E(r')\,dr'\right). \quad (1.5.\text{A8})$$

The nuclear reactions that occur within the radius r_0 fix the ratio C_+/C_- to take some value of order unity, which we will not need to calculate. We suppose further that $V(r)$ eventually decreases to zero for $r \to \infty$. Equation (1.5.A8) must break down when r approaches a radius r_E where $V(r_E) = E$, at which

$\kappa(r_E) = 0$. We take the potential barrier between r_0 and r_E to be sufficiently high and thick that
$$\int_{r_0}^{r_E} \kappa_E(r)\, dr \gg 1.$$
Then, for $r = r_E$, Eq. (1.5.A8) reads
$$\psi_E(r_E) \simeq C_+ \exp\left(+\int_{r_0}^{r_E} \kappa_E(r')\, dr'\right),$$
the other term in Eq. (1.5.A8) being negligible. For $r > r_E$ the function $\psi_E(r)$ oscillates, with little change in amplitude, so $|\psi_E(r_E)|$ is determined by the wave function representing the approach of the nuclei from a large separation. Thus the rate of nuclear reactions is suppressed by a barrier penetration factor
$$B(E) \simeq \left|\frac{C_+}{\psi_E(r_E)}\right|^2 = \exp\left(-2\int_{r_0}^{r_E} \kappa_E(r')\, dr'\right). \quad (1.5.\text{A}9)$$

For a Coulomb barrier, we have $V(r) = Z_1 Z_2 e^2 / r$, so, taking $r_0 \ll r_E$, we have
$$B(E) \simeq \exp\left[-2\int_0^{r_E} dr\, \sqrt{\frac{2\mu Z_1 Z_2 e^2}{\hbar^2}\left(\frac{1}{r} - \frac{1}{r_E}\right)}\right], \quad (1.5.\text{A}10)$$
where $r_E = Z_1 Z_2 e^2 / E$. To do this integral, we set $r = r_E u^2$, and use $\int_0^1 du\, \sqrt{1-u^2} = \pi/4$, so that
$$B(E) \simeq \exp\left[-\pi\sqrt{\frac{2\mu Z_1 Z_2 e^2 r_E}{\hbar^2}}\right] = \exp\left[-\pi Z_1 Z_2 e^2 \sqrt{\frac{2\mu}{\hbar^2 E}}\right], \quad (1.5.\text{A}11)$$
as was to be shown.

1.6 Relations among Observables: The Main Sequence

As we have seen in Section 1.3, we expect on very general grounds that stellar parameters such as radius, luminosity, central temperature, effective surface temperature, etc. all depend only on the star's mass, age, and initial chemical composition. This is why, when any pair of these parameters for a sample of stars in a cluster that all began at the same time with the same uniform chemical composition are plotted against each other, the values of these parameters will fall close to a one-dimensional curve, such as the Hertzsprung–Russell diagram comparing luminosity and effective surface temperature. But to find the *form* of these curves requires detailed physical assumptions and numerical calculation.

1.6 Relations among Observables: The Main Sequence

We shall see in this section that for stars that are still on the main sequence, burning hydrogen at their cores, it is possible to make a good estimate of the form of these curves using dimensional analysis, together with the assumption of power-law behavior for the rate per mass ϵ of nuclear energy generation and for the opacity κ:

$$\epsilon = \epsilon_1 \rho^\lambda (k_B T)^\nu, \qquad \kappa = \kappa_1 \rho^\alpha (k_B T)^\beta, \tag{1.6.1}$$

with κ_1 and ϵ_1, α, β, λ, and ν all constants assumed to depend only on chemical composition. (Section 1.4 found $\alpha = \beta = 0$ for Thomson scattering, and $\alpha = 1$ and $\beta = -7/2$ for free–free absorption. Section 1.5 found $\lambda = 1$ for the proton–proton chain and CNO cycle; $\nu \approx 5$ for the proton–proton chain and larger for the CNO cycle, and ν weakly dependent on temperature, with $\nu \propto T^{-1/3}$.) Our discussion in this section will be limited to stars in which thermal energy is transported only by radiation. In the following section we shall show that the presence of convective energy transport does not change our main conclusions.

With these assumptions, each stellar parameter will turn out to be dependent only on the star's mass M and a pair of quantities N_1 and N_2 that depend on chemical composition and fundamental physical constants. Since there are no dimensionless ratios among M, N_1, and N_2, any stellar parameter will be proportional to a product of powers of M, N_1, and N_2, with exponents fixed by dimensional analysis. This only works for stars on the main sequence whose chemical composition (on which κ_1, α, etc. depend) is still approximately uniform. For red giant stars whose stellar parameters also depend on the radius of the helium core, dimensional analysis is not enough. It is also not enough even if we assume that non-uniformities evolve from an initially uniform composition, because then stellar parameters depend on the *age* of the star, as well as on M, N_1, and N_2.

To carry out our dimensional analysis, we write Eqs. (1.3.3) and (1.3.4) in terms of ρ, $k_B T$, and $\mathcal{L}^* \equiv \mathcal{L}/\epsilon_1$:

$$\frac{d\mathcal{L}^*(r)}{dr} = 4\pi r^2 \rho^{\lambda+1}(r)\big(k_B T(r)\big)^\nu, \tag{1.6.2}$$

$$\frac{d\big(k_B T(r)\big)^4}{dr} = -3N_1 \rho^{\alpha+1}(r)\big(k_B T(r)\big)^\beta \frac{\mathcal{L}^*(r)}{4\pi r^2}, \tag{1.6.3}$$

where

$$N_1 \equiv \frac{\kappa_1 \epsilon_1 k_B^4}{ca}. \tag{1.6.4}$$

We will begin by assuming that the pressure p is dominated by gas pressure, as is the case for all but the most massive stars. (We will return at the end of this section to stars in which p is dominated by radiation pressure.) The pressure then is well approximated by the ideal gas law, $p = k_B T \rho / m_1 \mu$, where μ is the molecular weight and m_1 is the mass of unit atomic weight. Then Eqs. (1.3.1) and (1.3.2) are

$$\frac{d(\rho(r)k_B T(r))}{dr} = -N_2 \frac{\mathcal{M}(r)\rho(r)}{4\pi r^2}, \quad (1.6.5)$$

$$\frac{d\mathcal{M}(r)}{dr} = 4\pi r^2 \rho(r), \quad (1.6.6)$$

where

$$N_2 \equiv 4\pi G m_1 \mu. \quad (1.6.7)$$

For uniform chemical composition, the stellar parameters R, $L^* \equiv L/\epsilon_1$, $\rho(0)$, $k_B T(0)$, etc. can depend only on N_1, N_2, and M.

Next we must work out the dimensionalities of N_1 and N_2 in powers of length, time, and mass. We note that the energy production rate per mass has dimensions

$$[\epsilon] = [\text{energy}][\text{mass}]^{-1}[\text{time}]^{-1} = [\text{velocity}]^2[\text{time}]^{-1} = [\text{length}]^2[\text{time}]^{-3},$$

so

$$[\epsilon_1] = [\text{length}]^2[\text{time}]^{-3}[\text{mass}/\text{length}^3]^{-\lambda}[\text{energy}]^{-\nu}$$
$$= [\text{length}]^{2+3\lambda-2\nu}[\text{time}]^{-3+2\nu}[\text{mass}]^{-\lambda-\nu}.$$

Also, since $1/\kappa\rho$ is the mean free path, the opacity has dimensions $[\kappa] = [\text{length}]^{-1}/[\text{mass}/\text{length}^3]$, so

$$[\kappa_1] = [\text{length}]^{-1}[\text{mass}/\text{length}^3]^{-1-\alpha}[\text{energy}]^{-\beta}$$
$$= [\text{length}]^{2+3\alpha-2\beta}[\text{time}]^{2\beta}[\text{mass}]^{-1-\alpha-\beta}.$$

Finally,

$$[ca/k_B^4] = [\text{energy}][\text{time}]^{-1}[\text{area}]^{-1}[\text{energy}]^{-4}$$
$$= [\text{energy}]^{-3}[\text{length}]^{-2}[\text{time}]^{-1}$$
$$= [\text{length}]^{-8}[\text{time}]^5[\text{mass}]^{-3}.$$

Thus

$$[N_1] = [\text{length}]^{12+3\lambda-2\nu+3\alpha-2\beta}[\text{time}]^{-8+2\nu+2\beta}[\text{mass}]^{2-\lambda-\nu-\alpha-\beta} \quad (1.6.8)$$

and

$$[N_2] = [G][\text{mass}] = [\text{velocity}]^2[\text{length}] = [\text{length}]^3[\text{time}]^{-2}. \quad (1.6.9)$$

To calculate the stellar radius R, we ask what product of form $M^A N_1^{A_1} N_2^{A_2}$ has the dimensions of length. Setting the numbers of powers of length, time, and mass in this product respectively equal to $+1$, 0, and 0, we find

1.6 Relations among Observables: The Main Sequence

powers of length : $1 = (12 + 3\lambda - 2\nu + 3\alpha - 2\beta)A_1 + 3A_2,$ (1.6.10)
powers of time : $0 = (-8 + 2\nu + 2\beta)A_1 - 2A_2,$ (1.6.11)
powers of mass : $0 = A + (2 - \lambda - \nu - \alpha - \beta)A_1.$ (1.6.12)

Using Eq. (1.6.11) to eliminate A_2 in Eq. (1.6.10) gives A_1; Eq. (1.6.11) then gives A_2; and using this in Eq. (1.6.12) gives A. In this way we find

$$A = \frac{-2 + \lambda + \nu + \alpha + \beta}{3\lambda + \nu + 3\alpha + \beta},$$ (1.6.13)

$$A_1 = \frac{1}{3\lambda + \nu + 3\alpha + \beta},$$ (1.6.14)

$$A_2 = \frac{-4 + \nu + \beta}{3\lambda + \nu + 3\alpha + \beta},$$ (1.6.15)

and so

$$R \cong M^A N_1^{A_1} N_2^{A_2},$$ (1.6.16)

with A, A_1, and A_2 given by Eqs. (1.6.13)–(1.6.15). (Here we use $\cong$ to mean "proportional to, and since there are no very large or very small dimensionless constants in the differential equations, also roughly equal to.")

Likewise, the luminosity has dimensions

$$[L] = [\text{energy}]/[\text{time}] = [\text{length}]^2 [\text{time}]^{-3} [\text{mass}],$$

so $L^* \equiv L/\epsilon_1$ has dimensions

$$[L^*] = [\text{length}]^{-3\lambda + 2\nu} [\text{time}]^{-2\nu} [\text{mass}]^{1+\lambda+\nu}.$$

Following the same procedure as above for R, we find that the unique product of powers of M, N_1, and N_2 that has the same dimensionality as L^* is

$$L^* \cong M^B N_1^{B_1} N_2^{B_2},$$

where

$$B = \frac{(1 + \lambda + \nu)(3\alpha + \beta) + (3 - \alpha - \beta)(3\lambda + \nu)}{3\lambda + \nu + 3\alpha + \beta},$$ (1.6.17)

$$B_1 = -\frac{3\lambda + \nu}{3\lambda + \nu + 3\alpha + \beta},$$ (1.6.18)

$$B_2 = \frac{\nu(3\alpha + \beta) + (4 - \beta)(3\lambda + \nu)}{3\lambda + \nu + 3\alpha + \beta}.$$ (1.6.19)

We conclude then that

$$L = \epsilon_1 L^* \cong \epsilon_1 M^B N_1^{B_1} N_2^{B_2}.$$ (1.6.20)

The same reasoning can be applied to other quantities, such as the temperature at the center of the star. The only combination of N_1, N_2, and M that has the same dimensions as $k_B T$ is $M^C N_1^{C_1} N_2^{C_2}$, where

$$C = 2\frac{\lambda + \alpha + 1}{3\lambda + \nu + 3\alpha + \beta}, \tag{1.6.21}$$

$$C_1 = -\frac{1}{3\lambda + \nu + 3\alpha + \beta}, \tag{1.6.22}$$

$$C_2 = 1 + 4C_1, \tag{1.6.23}$$

so we conclude that the central temperature is

$$k_B T(0) \cong M^C N_1^{C_1} N_2^{C_2}. \tag{1.6.24}$$

At this point the reader may be wondering why the central temperatures of stars are so different from their effective surface temperatures, despite their having the same dimensionality. For instance, the effective surface temperature of the Sun is measured as $T_{\text{eff},\odot} = 5,800$ K, while detailed solar models give the central temperature of the Sun as $T_\odot(0) = 1.36 \times 10^7$ K etc., larger by a factor 2,340. The answer is that, while the central temperature depends only on M, N_1, and N_2, this is not true of the effective surface temperature, which is defined by the requirement $L = 4\pi R^2 \sigma T_{\text{eff}}^4$, or in other words,

$$k_B T_{\text{eff}} \equiv [k_B^4 L / 4\pi \sigma R^2]^{1/4} = [L k_B^4 / \pi a c R^2]^{1/4} = [N_1 L^* / \pi R^2 \kappa_1]^{1/4}. \tag{1.6.25}$$

This can be written as the product

$$T_{\text{eff}} = \tau_0^{-1/4} T_0, \tag{1.6.26}$$

where τ_0 is the dimensionless quantity

$$\tau_0 = R \kappa_1 [M/R^3]^{1+\alpha} [k_B T(0)]^\beta,$$

and T_0 has the dimensions of temperature,

$$k_B T_0 = \left[R[M/R^3]^{1+\alpha} [k_B T(0)]^\beta N_1 L^* / \pi R^2 \right]^{1/4}.$$

Since T_0 and $T(0)$ depend only on M, N_1, and N_2, and have the same dimensionality, we expect them to be equal, up to factors of order unity. So from Eq. (1.6.26) we expect that

$$T(0)/T_{\text{eff}} \approx \tau_0^{1/4}. \tag{1.6.27}$$

On the other hand, τ_0 is the value that the optical depth of the center of the star would have if the density and temperature had the uniform values M/R^3 and $T(0)$, which is much greater than unity because the star is optically thick. For instance, if we take the Sun to be completely ionized hydrogen and take its opacity to be entirely due to Thomson scattering, then, as shown in Section 1.4,

1.6 Relations among Observables: The Main Sequence

$\kappa \simeq 0.4$ cm^2/g, so for a uniform density $\approx M/R_\odot^3$ the optical depth of the center of the Sun is $\tau_0 \approx R_\odot \kappa M_\odot / R_\odot^3 = 1.6 \times 10^{11}$. Our estimate (1.6.27) then suggests that $T(0)/T_{\rm eff} \simeq 630$, not very different from the actual ratio 2,340 cited above.

We are now in a position to find the shape of the famous Hertzsprung–Russell relation between effective surface temperature and luminosity for stars on the main sequence. From the definition (1.6.25) and our results that $R \propto M^A$ and $L \propto M^B$, we find the mass dependence of the effective surface temperature

$$T_{\rm eff} \propto M^{[B-2A]/4}. \tag{1.6.28}$$

Therefore, eliminating M from our results for L and $T_{\rm eff}$, we can express the Hertzsprung–Russell relation as a power law:

$$L \propto T_{\rm eff}^H \tag{1.6.29}$$

with exponent

$$H = \frac{4B}{B-2A} = 4\left[1 - 2\frac{-2+\lambda+\nu+\alpha+\beta}{(1+\lambda+\nu)(3\alpha+\beta)+(3-\alpha-\beta)(3\lambda+\nu)}\right]^{-1}. \tag{1.6.30}$$

The estimate of H is simplest for stars on the upper part of the main sequence, whose high temperature means that opacity is dominated by Thomson scattering, for which $\alpha = \beta = 0$. For both the proton–proton chain and the CNO cycle $\lambda = 1$, so leaving ν as a free parameter, the Hertzsprung–Russell exponent is

$$H = \frac{12(3+\nu)}{11+\nu}. \tag{1.6.31}$$

In all cases ν is positive-definite and $3.27 < H < 12$. More specifically, for the proton–proton chain and CNO cycle we have roughly $\nu \simeq 5$ and $\nu \simeq 15$, for which respectively $H \simeq 6$ and $H \simeq 8.3$. The comparison with observation is complicated by the fact that, although it is straightforward to measure L for any star whose distance is known (or to measure ratios of values of L for a cluster of stars that are all at the same distance), it is difficult to obtain a precise value for $T_{\rm eff}$ from observations of colors or spectral lines. From one graph[31] of L versus $T_{\rm eff}$ for a large sample of stars with masses between 2 and 10 solar masses, I estimate that $H \simeq 7$.

The problems associated with the measurement of effective surface temperature can be avoided by considering the class of eclipsing binary stars, for which accurate values of R and M can be found from the analysis of the time-dependence of luminosities and Doppler shifts.[32] It is particularly revealing to consider the relation between luminosity and mass for stars, such as those on the

---
[31] F. LeBlanc, *Introduction to Stellar Astrophysics* (John Wiley & Sons, Chichester, 2010), p. 27.
[32] J. Andersen, *Astron. Astrophys. Rev.* **3**, 91 (1991).

upper part of the main sequence, whose opacity is due to Thomson scattering, for which $\alpha = \beta = 0$. For these stars Eqs. (1.6.17)–(1.6.19) give $B = 3$, $B_1 = -1$, and $B_2 = 4$, so here Eq. (1.6.20) reads

$$L \cong \epsilon_1 M^3 N_1^{-1} N_2^4 = \frac{ca(4\pi G m_1 \mu)^4}{\kappa_1 k_B^4} M^3. \tag{1.6.32}$$

It is striking that this result is entirely independent of the parameters ϵ_1, λ, and ν characterizing the mechanism for nuclear energy generation. One suspects that for $\alpha = \beta = 0$ this result is even independent of the assumption that the rate of energy generation per mass is proportional to a product of powers of density and temperature, but I have not been able to prove this.

The data on eclipsing binaries cited by Andersen shows that for $2 \leq M/M_\odot \leq 20$, binaries have $L \propto M^{3.6}$. Another survey[33] shows that bright stars have $L \propto M^{4.0}$, while dimmer stars have $L \propto M^{2.76}$. Stars on the upper part of the main sequence have[34] $L \propto M^{3.5}$. Given the limited statistics from eclipsing binaries and the oversimplification in our assumption of an opacity entirely due to Thomson scattering, the discrepancies among these measured exponents – 3.6, 4.0, 2.76, and 3.5 – and with our result that $L \propto M^3$ – are not surprising.

The luminosity–mass relation provides insight regarding the scale of time over which stars of various mass evolve. The fusion $4\,^1\mathrm{H} \to {}^4\mathrm{He}$ yields 6.5 MeV per proton, so the energy per mass available from hydrogen burning is

6.5 MeV/p $\times\ 1.602 \times 10^{-5}$ erg/MeV$/1.672 \times 10^{-24}$ g/p $= 6.23 \times 10^{19}$ erg/g.

The Sun has mass $M_\odot = 1.939 \times 10^{33}$ g, but initially only 75% was hydrogen, so the energy available is

$$E_\mathrm{H} = 0.75 f \times 1.939 \times 10^{33} \left(\frac{M}{M_\odot}\right) \mathrm{g} \times 6.23 \times 10^{19}\ \mathrm{erg/g}$$

$$= 0.93 \times 10^{53} f \left(\frac{M}{M_\odot}\right) \mathrm{erg},$$

where f is the fraction of the Sun's hydrogen that becomes sufficiently hot to initiate nuclear reactions. The Sun has luminosity $L_\odot = 3.845 \times 10^{33}$ erg/sec, so a star of mass M and luminosity L could go on burning hydrogen for a time

$$E_\mathrm{H}/L \simeq 7.6 \times 10^{11} f \frac{M/M_\odot}{L/L_\odot}\ \text{years}.$$

The main sequence duration of the Sun is commonly estimated as 10^{10} years, corresponding to $f \simeq 0.013$, a not unreasonable value. Even with an efficiency

---

[33] Cited by J. P. Cox and R. T. Giuli, *Principles of Stellar Structure* (Gordon and Breach, New York, 1968), p. 15.

[34] C. J. Hansen, S. D. Kawaler, and V. Trimble, *Stellar Interiors: Physical Principles, Structure and Evolution*, 2nd edn. (Springer, New York, 2004), p. 28.

1.6 Relations among Observables: The Main Sequence

this small, the solar main sequence lifetime is much longer than the Kelvin time 10^7 years over which the Sun could go on shining without nuclear reactions, and it is not much less than the present age 1.37×10^{10} years of the big bang. But with our analytic estimate $L \propto M^3$ and the same hydrogen burning efficiency, for $M = 100 M_\odot$ the main sequence duration would be only 10^6 years, while with the empirical relation $L \propto M^{3.5}$ the main sequence lifetime would be 10^5 years.

Finally, consider the relation between stellar radii and masses. Recall that $R \propto M^A$, and for $\alpha = \beta = 0$ and $\lambda = 1$, Eq. (1.6.13) gives

$$A = \frac{-1+\nu}{3+\nu}.$$

If for the CNO cycle we take $\nu = 15$, then $R \propto M^{0.78}$. Data[35] for stars with masses between 5 and 20 solar masses give $R \propto M^{0.78}$, while other data[36] for stars on the upper part of the main sequence indicate that $R \propto M^{0.75}$. This is a very satisfactory confirmation of the results of dimensional analysis.

* * * * *

In closing, we return to the case in which the pressure is dominated by radiation rather than hot gas. Here $p = aT^4/3$, so Eqs. (1.6.5) and (1.6.7) are replaced with

$$\frac{d(k_B T(r))^4}{dr} = -3N_2' \frac{\mathcal{M}(r) \rho(r)}{4\pi r^2},$$

and

$$N_2' \equiv 4\pi G k_B^4 / a,$$

while there is no change in Eqs. (1.6.1)–(1.6.4) or (1.6.6). Now stellar parameters R, $L^* \equiv L/\epsilon_1$, $\rho(0)$, $k_B T(0)$, etc. depend only on N_1, N_2', and M. Note that N_2' has the dimensions of $G[\text{energy}]^4/[\text{energy}/\text{volume}]$, or

$$[N_2'] = [G][\text{energy}]^4/[\text{energy}/\text{volume}] = [G][\text{energy}]^3[\text{volume}]$$
$$= [\text{length}]^{12}[\text{time}]^{-8}[\text{mass}]^2. \qquad (1.6.33)$$

Here again there is a remarkably general simple relation between luminosity and mass in the case where opacity is dominated by Thomson scattering. Recall that L has dimensions

$$[L] = [\text{energy}]/[\text{time}] = [\text{length}]^2[\text{time}]^{-3}[\text{mass}],$$

---

[35] Cited by A. Weiss, W. Hillebrandt, H.-C. Thomas, and H. Ritter, *Cox and Giuli's Principles of Stellar Structure*, 2nd edn. (Cambridge Scientific Publishers, Cambridge, 2004), p. 10.

[36] Cited by C. J. Hansen, S. D. Kawaler, and V. Trimble, *Stellar Interiors: Physical Principles, Structure and Evolution*, 2nd edn. (Springer, New York, 2004), p. 28.

so $L^* \equiv L/\epsilon_1$ has dimensions

$$[L^*] = [\text{length}]^{-3\lambda+2\nu}[\text{time}]^{-2\nu}[\text{mass}]^{1+\lambda+\nu}.$$

The only combination of N_1, N_2', and M that has the same dimensions as L^* is $M^{B'}N_1^{B_1'}N_2'^{B_2'}$, where

$$B' = 1 + \frac{(\lambda+\nu/2)(3\alpha+\beta) - (\alpha+\beta/2)(3\lambda+\nu)}{3\lambda+\nu+3\alpha+\beta}, \tag{1.6.34}$$

$$B_1' = -\frac{3\lambda+\nu}{3\lambda+\nu+3\alpha+\beta}, \tag{1.6.35}$$

$$B_2' = \frac{\nu}{4} + B_1'\left(-1 + \frac{\nu}{4} + \frac{\beta}{4}\right). \tag{1.6.36}$$

Hence

$$L = \epsilon_1 L^* \cong \epsilon_1 M^{B'} N_1^{B_1'} N_2'^{B_2'}. \tag{1.6.37}$$

Equations (1.6.25)–(1.6.27) with $\alpha = \beta = 0$ give $B' = 1$, $B_1' = -1$, $B_2' = 1$, so Eq. (1.6.37) gives

$$L \cong \epsilon_1 M N_1^{-1} N_2'^{1} = \frac{4\pi G c}{\kappa_1} M, \tag{1.6.38}$$

for any values of λ and ν. This may be compared with the result given at the end of Section 1.3, that in the absence of gas pressure

$$L = \frac{4\pi G c M}{\kappa(R)}. \tag{1.6.39}$$

This result was derived with no assumptions regarding the dependence of opacity or nuclear energy generation on temperature and density. If we assume that the opacity is independent of temperature and density, as it is for Thomson scattering, then $\kappa(R)$ is the same as what in this section we have called κ_1, so Eq. (1.6.38) is the same as Eq. (1.6.39) and dimensional analysis is not needed. Here $\cong$ actually means $=$. But, for a more general dependence of opacity on temperature and density, of the form (1.6.1), though Eq. (1.6.39) is still valid in the absence of gas pressure, the opacity $\kappa(R)$ at the surface is different from κ_1, the relation depending on the profile of density and pressure throughout the star and hence on the stellar mass, so here dimensional analysis comes in handy in finding the luminosity–mass relation.

1.7 Convection

The regime of radiative energy transport discussed in the previous sections, and more generally any smooth model of stellar structure, may not be stable against the onset of convection. Bits of stellar material may separate from their surroundings, and rise or fall, like eddies in a heated pot of water.

Suppose that a small element of stellar fluid happens to move upward from r to $r + dr$. The balance of forces at the surface of the element will cause the pressure inside to change, from $p(r)$ to the ambient pressure $p(r) + p'(r) dr$ at its new location. The density and temperature will also change, but not to the new ambient density and temperature. Since heat conduction is generally very slow in stars, it is reasonable to suppose that the process is adiabatic, with no heat flowing into or out of the fluid element. Then the density and temperature will be some definite function of the pressure (in general depending on initial conditions), and in particular the new density will be

$$\rho(r) + \left[\frac{\partial \rho(p)}{\partial p}\right]_{p=p(r)} p'(r) \, dr, \tag{1.7.1}$$

in which we adopt the convention that a partial derivative in square brackets is to be calculated assuming that variations are adiabatic – that is, with changes in pressure, the temperature and density vary in such a way that no heat flows into or out of the fluid element. If this new density is greater than the ambient density $\rho(r) + \rho'(r) \, dr$ at the new position, then the fluid element will sink back toward its original position, and the initial configuration will be stable. Thus the condition for stability against upward motion is

$$\left[\frac{\partial \rho(p)}{\partial p}\right]_{p=p(r)} p'(r) > \rho'(r). \tag{1.7.2}$$

Similarly, if the blob density (1.7.1) is less than the new ambient density $\rho(r) + \rho'(r) \, dr$ then the fluid element will float upward, so we then have stability against downward motion. Since for downward motion dr is negative, the stability condition is again (1.7.2).

On the other hand, if the left-hand side of Eq. (1.7.2) is less than the right-hand side we have an exponentially growing instability, whereas if the two sides are equal we have instability against a steady drift upwards or downwards.

Under conditions of convective stability, the r-derivative of the temperature is given by the equation (1.3.4) of radiative energy transport, while the r-derivative of the pressure is given by the equation (1.3.1) of hydrostatic equilibrium, so it is convenient (and conventional) to rewrite the equation (1.7.2) of convective stability in terms of temperature and pressure rather than density and pressure. For this purpose, we need the ideal gas law

$$\rho = mp/k_B T,$$

where m is the mass of the gas particles, whose value will not concern us. It follows then that

$$\left[\frac{\partial \rho}{\partial p}\right] = \frac{\rho}{p} - \frac{\rho}{T}\left[\frac{\partial T}{\partial p}\right] = \frac{\rho}{p}\left\{1 - \left[\frac{\partial \ln T}{\partial \ln p}\right]\right\}$$

(the square brackets again indicating adiabatic variations), and
$$\rho' = \frac{\rho p'}{p} - \frac{T'\rho}{T},$$
so the quantity appearing in the stability condition (1.7.2) can be written
$$\left[\frac{\partial \rho}{\partial p}\right]_{p=p(r)} p'(r) - \rho'(r) = -\frac{p'(r)\rho(r)}{p(r)}\left(\nabla_{\rm ad}(r) - \nabla(r)\right),$$
where $\nabla_{\rm ad}$ is the value of $\partial \ln T / \partial \ln p$ for adiabatic variations,
$$\nabla_{\rm ad}(r) \equiv \left[\frac{\partial \ln T(p)}{\partial \ln p}\right]_{p=p(r)}, \qquad (1.7.3)$$
and $\nabla(r)$ is the actual value of this derivative in the star:
$$\nabla(r) \equiv \frac{T'(r)/T(r)}{p'(r)/p(r)}. \qquad (1.7.4)$$
Since the quantity $p'\rho/p$ is everywhere negative, the condition (1.7.2) for convective stability is just
$$\nabla(r) < \nabla_{\rm ad}(r). \qquad (1.7.5)$$
Using Eqs. (1.3.1) and (1.3.4), we have
$$\nabla(r) = \frac{3\kappa(r)\mathcal{L}(r)p(r)}{16\pi ca\, T^4(r)G\mathcal{M}(r)}. \qquad (1.7.6)$$
It is instructive to write the stability condition $\nabla(r) < \nabla_{\rm ad}(r)$ as a limit on the rate of energy flow through a sphere of radius r that can be carried stably by radiation:
$$\mathcal{L}(r) < 4\nabla_{\rm ad}(r)\left(\frac{p_{\rm rad}(r)}{p(r)}\right)\mathcal{L}_{\rm Edd}(r), \qquad (1.7.7)$$
where $p_{\rm rad}(r)$ is the radiation pressure $aT(r)^4/3$ and $\mathcal{L}_{\rm Edd}(r)$ is the Eddington limit $4\pi Gc\mathcal{M}(r)/\kappa(r)$. As we saw at the end of Section 1.3, $\mathcal{L}(r)$ must in any case be less than $\mathcal{L}_{\rm Edd}(r)$ in order for radiation not to overcome gravitational attraction and tear the star apart. We will see that $4\nabla_{\rm ad}$ is never very different from unity, so for ordinary stars, for which radiation pressure is much less than gas pressure, stability against convection requires that $\mathcal{L}(r)$ must be not just less but very much less than the Eddington limit $\mathcal{L}_{\rm Edd}(r)$.

To calculate $\nabla_{\rm ad}$ we make use of the conservation of energy and mass. (For relativistic theories, in which mass is not conserved, we use baryon number instead.) We take $\mathcal{E}$ as the thermal energy density, excluding the energy associated with rest masses, so the thermal energy per gram is $\mathcal{E}/\rho$. When the volume per gram $1/\rho$ of stellar material increases by a small amount $\delta(1/\rho)$ (which of course is negative for decreasing volume), the work per gram that is done against the ambient pressure p is $p\,\delta(1/\rho)$, so in the absence of heat flow the conservation of energy requires that

$$\delta(\mathcal{E}/\rho) + p\,\delta(1/\rho) = 0. \qquad (1.7.8)$$

As already mentioned in Section 1.1, for a wide variety of stellar material $\mathcal{E}$ is proportional to p, a relation conventionally written as

$$\mathcal{E} = \frac{p}{\Gamma - 1}. \qquad (1.7.9)$$

(This is sometimes written as $\mathcal{E} = np$, where $n \equiv 1/(\Gamma - 1)$.) Using Eq. (1.7.9) in Eq. (1.7.8), the adiabatic energy conservation condition becomes

$$\Gamma p\,\delta(1/\rho) + (1/\rho)\,\delta p = 0,$$

or in other words

$$\delta(p/\rho^\Gamma) = 0. \qquad (1.7.10)$$

The adiabatic partial derivative in Eq. (1.7.2) is then

$$\left[\frac{\partial \rho(p)}{\partial p}\right] = \frac{\rho}{\Gamma p}, \qquad (1.7.11)$$

and the stability condition (1.7.2) is then just the condition that

$$\frac{\rho(r)\,p'(r)}{\Gamma p(r)} > \rho'(r),$$

or, multiplying by the positive quantity $\Gamma/\rho(r)$,

$$\frac{p'(r)}{p(r)} > \frac{\Gamma \rho'(r)}{\rho(r)}. \qquad (1.7.12)$$

(The difference between the left-hand and right-hand sides of this inequality is a quantity known as the *Schwarzschild discriminant*.) Hence stability requires that $p(r)/\rho^\Gamma(r)$ increases with r. Where this is not the case, convection occurs.

For an ideal gas, with p proportional to ρT, we have p/ρ^Γ proportional to $T^\Gamma/p^{(\Gamma-1)}$, so for adiabatic variations $T \propto p^{(\Gamma-1)/\Gamma}$, and the quantity (1.7.3) is the constant

$$\nabla_{\text{ad}} = 1 - 1/\Gamma. \qquad (1.7.13)$$

This is the value we must use in the stability criterion (1.7.5).

For a monatomic ideal gas of atoms at temperature T the equipartition of energy gives a thermal energy per atom $3k_\text{B}T/2$, so with $\rho/m_1\mu$ atoms per volume (where μ is the atomic weight and m_1 the mass for unit atomic weight), the thermal energy per volume is $\mathcal{E} = 3k_\text{B}T\rho/2\mu m_1$, as compared with a pressure given by the ideal gas law as $p = k_\text{B}T\rho/\mu m_1$, so here Eq. (1.7.9) is satisfied with $\Gamma = 1 + 2/3 = 5/3$, and $\nabla_{\text{ad}} = 2/5$.

Matters are not always so simple, even in ordinary stars. For instance in the Sun, as we go inwards from just below the surface to $r \simeq 0.8R_\odot$, the increasing temperature goes first to ionizing atomic hydrogen (which takes 13.6 eV per atom), then to singly ionizing atomic helium (24.6 eV per atom), and then to completely ionizing singly ionized helium (54.4 eV per ion), rather than to increasing thermal velocities and pressure. Since $\partial \mathcal{E}/\partial p$ is thus effectively greater than 3/2, the effective value of Γ is less than 5/3, and ∇_{ad} is less than 2/5. In the outer layers of the Sun, from just below the surface down to $r \simeq 0.8 R_\odot$, the effective value of ∇_{ad} is approximately 0.15.[37] Elsewhere in the Sun, ∇_{ad} is close to the nominal value 2/5.

Energy density is proportional to pressure also if the thermal energy and pressure are both dominated by relativistic particles, such as fast electrons in high-mass white dwarfs or photons in supermassive stars. In such cases we have $p = \mathcal{E}/3$, so Eqs. (1.7.9) and (1.7.13) are satisfied with $\Gamma = 4/3$ and $\nabla_{\mathrm{ad}} = 1/4$.

Now suppose that, in some part of a star, the condition (1.7.2) for convective stability is not satisfied, but rather

$$\left[\frac{\partial \rho(p)}{\partial p}\right]_{p=p(r)} p'(r) < \rho'(r), \tag{1.7.14}$$

or equivalently,

$$\nabla(r) > \nabla_{\mathrm{ad}}(r). \tag{1.7.15}$$

(This is the case in the Sun from just below the surface, at a depth where $p \approx 10^5$ dyne/cm^2, down to $r \simeq 0.7 R_\odot$, where $p \approx 10^{13.5}$ dyne/cm^2.) As we have seen, in this case a blob of stellar fluid that happens to move upwards or downwards will become respectively lighter or heavier than the same volume of ambient fluid along its path, and hence will tend to keep moving in the same direction. The pressure in the blob remains the same as the ambient pressure along its path, so if the energy per volume $\mathcal{E}$ depends only on the pressure, it too remains the same in the blob as in the fluid along its path, but since the mass density ρ in the blob becomes less or greater than in the fluid along its path for a blob going upwards or downwards, the energy per mass $\mathcal{E}(p)/\rho$ becomes respectively greater or less than in the fluid along its path. Specifically, after the blob travels a distance δr, the difference between its density and the density of the surrounding material will be

$$\delta\rho = \left[\left[\frac{\partial \rho(p)}{\partial p}\right]_{p=p(r)} p'(r) - \rho'(r)\right] \delta r,$$

so the difference between the thermal energy per mass of the blob and of the surrounding material will be

---

[37] Numerical results for ∇ and ∇_{ad} here and below are taken from Figure 29.4 of R. Kippenhahn and A. Weigert, *Stellar Structure and Evolution* (Springer-Verlag, Berlin, 1990). Other solar parameters are taken from C. W. Allen, *Astrophysical Quantities* (Athlone Press, London, 1955).

$$\delta\left(\frac{\mathcal{E}}{\rho}\right) = -\mathcal{E}\delta\rho/\rho^2 = \frac{\mathcal{E}(r)}{\rho^2(r)}\left[\rho'(r) - \left[\frac{\partial\rho(p)}{\partial p}\right]_{p=p(r)} p'(r)\right]\delta r. \quad (1.7.16)$$

According to the condition (1.7.14) for convection to occur, the change (1.7.16) in energy per mass of the blob will be positive or negative for outward or inward motion, respectively. Eventually the blob will dissolve into the ambient material, heating the ambient matter above if the blob has gone upward and cooling the matter below if the blob has gone downward. The succession of blobs going up and down thus leads to a flow of heat energy outward through the star.

The convective transport of energy forces a clarification of notation. From now on, we refer to the rate of energy transport outward through a sphere of radius r by radiation and convection as $\mathcal{L}_{\text{rad}}(r)$ and $\mathcal{L}_{\text{conv}}(r)$, respectively, while the total rate of energy transport is

$$\mathcal{L}_{\text{rad}}(r) + \mathcal{L}_{\text{conv}}(r) \equiv \mathcal{L}_{\text{tot}}(r).$$

The equation (1.3.3) of energy conservation refers of course to the total energy transport rate

$$\frac{d\mathcal{L}_{\text{tot}}(r)}{dr} = 4\pi r^2 \epsilon(r)\rho(r), \quad (1.7.17)$$

while it is $\mathcal{L}_{\text{rad}}(r)$ that controls variations in temperature through Eq. (1.3.4), which we now write as

$$\frac{dT(r)}{dr} = -\frac{3\kappa(r)\rho(r)}{4caT^3(r)}\frac{\mathcal{L}_{\text{rad}}(r)}{4\pi r^2}. \quad (1.7.18)$$

Thus, in the presence of convection, Eq. (1.7.6) refers to the radiative energy transport rate, not the total rate:

$$\nabla(r) = \frac{3\kappa(r)\mathcal{L}_{\text{rad}}(r)p(r)}{16\pi ca\, T^4(r)GM(r)}. \quad (1.7.19)$$

Often one defines a quantity $\nabla_{\text{rad}}(r)$ as what $\nabla(r)$ would be if energy were transported entirely by radiation:

$$\nabla_{\text{rad}}(r) \equiv \frac{3\kappa(r)\mathcal{L}_{\text{tot}}(r)p(r)}{16\pi ca\, T^4(r)GM(r)}. \quad (1.7.20)$$

Since convection carries some energy, the presence of convection means that $\nabla(r)$ is less than $\nabla_{\text{rad}}(r)$ (often much less), as well as greater than ∇_{ad}.

Finding $\mathcal{L}_{\text{tot}}(r)$ is relatively easy. Equation (1.7.17) tells us that outside a central core where nuclear reactions occur, $\mathcal{L}_{\text{tot}}(r)$ is a constant, and hence is equal to the star's luminosity L. But in order to use Eq. (1.7.18) to calculate the variation in the star's temperature, we need to find $\mathcal{L}_{\text{rad}}(r)$, which is not so easy. Instead we can often simply assume that in convective zones $\nabla(r) \simeq \nabla_{\text{ad}}(r)$, so that the temperature varies in such a way as to keep the pressure simply proportional to ρ^Γ.

To see when this is likely to be the case, it is usual to calculate the convective energy flux employing a radical approximation. One assumes that the dissolution of each blob occurs after it has traveled a distance $\ell(r)$, known as the *mixing length*. (The mixing length at radius r is usually taken to be of the same order of magnitude as the scale height of the stellar fluid at that position, the radial distance in which density, pressure, etc. change appreciably, but it is difficult to justify this guess, and even more difficult to do better.) We assume that the whole mass of the star is involved in this convection, so the energy per time transported by convection through a sphere of radius r is the quantity (1.7.16) (with δr replaced with ℓ) times the mass $4\pi r^2 \rho(r)\ell(r)$ in a shell of thickness $\ell(r)$ divided by the time $\approx \ell(r)/u(r)$ that it takes blobs to pass through this shell,

$$\mathcal{L}_{\text{conv}}(r) \approx 4\pi r^2 u(r) \frac{\mathcal{E}(r)}{\rho(r)} \left(\rho'(r) - \left[\frac{\partial \rho(p)}{\partial p} \right]_{p=p(r)} p'(r) \right) \ell(r), \quad (1.7.21)$$

where $u(r)$ is a typical blob velocity. To estimate $u(r)$, we note that the buoyant force on a blob of volume V is the acceleration of gravity $g = G\mathcal{M}/r^2 = |p'/\rho|$ times the mass ρV of the ambient material with the same volume V minus the mass $(\rho + \delta\rho)V$ of the blob.[38] To first order the acceleration of the blob is this force divided by ρV, which after traveling a distance $\ell(r)$ is

$$a = \left| \frac{p'(r)}{\rho^2(r)} \right| \left(\rho'(r) - \left[\frac{\partial \rho(p)}{\partial p} \right]_{p=p(r)} p'(r) \right) \ell(r).$$

The average velocity over this time is then of the order $u \approx \sqrt{a\ell}$, or

$$u(r) \approx \left| \frac{p'(r)}{\rho^2(r)} \right|^{1/2} \left(\rho'(r) - \left[\frac{\partial \rho(p)}{\partial p} \right]_{p=p(r)} p'(r) \right)^{1/2} \ell(r). \quad (1.7.22)$$

Together with Eq. (1.7.21), this gives the energy per time transported by convection through a sphere of radius r as

---

[38] A proof of the classic result that the buoyant force on a submerged body equals the weight of the fluid the body displaces was given by Archimedes, "On Floating Bodies," in *The Complete Works of Archimedes*, trans. T. L. Heath (Cambridge University Press, Cambridge, 1897). He compared two columns of fluids. In one, the submerged body is held down by a piston, while in the other, with the same horizontal cross section as the submerged body, the fluid is undisturbed. In order for the fluid to be at rest the force pressing down at the base of the two columns must be the same, so the buoyancy, which equals the force exerted by the piston, plus the weight of the submerged body, plus the weight of the column of fluid less the weight of the fluid displaced by the body, must equal the weight of the fluid in the undisturbed column, which does include the weight of the fluid displaced by the body. The same result can be derived more directly by modern methods. The integral of the pressure force on the surface of the displaced body is related by Gauss's theorem to the integral of the pressure gradient over the displaced volume, which according to the equation of hydrostatic equilibrium equals the weight of the displaced fluid.

$$\mathcal{L}_{\text{conv}}(r) \approx 4\pi r^2 \left|\frac{p'(r)}{\rho^2(r)}\right|^{1/2} \frac{\mathcal{E}(r)}{\rho(r)} \left(\rho'(r) - \left[\frac{\partial \rho(p)}{\partial p}\right]_{p=p(r)} p'(r)\right)^{3/2} \ell^2(r). \tag{1.7.23}$$

In the same way as in our derivation of the condition (1.7.5) for convective stability, for ideal gases we can rewrite Eq. (1.7.23) as

$$\mathcal{L}_{\text{conv}}(r) \approx \mathcal{L}_0(r)\left(\nabla(r) - \nabla_{\text{ad}}(r)\right)^{3/2}, \tag{1.7.24}$$

where

$$\mathcal{L}_0(r) \equiv 4\pi r^2 \frac{p'^2(r)\mathcal{E}(r)\ell^2(r)}{p^{3/2}(r)\rho^{1/2}(r)}. \tag{1.7.25}$$

We say that convection is efficient at r if the coefficient $\mathcal{L}_0(r)$ is much larger than the luminosity L. This is often the case. Where the mixing length $\ell(r)$ is half the pressure scale height, and $\mathcal{E} = 3p/2$, we have

$$\mathcal{L}_0 \approx (3/2)\pi r^2 p^{3/2}(r)/\rho^{1/2}(r).$$

In the Sun at $r = 0.8 R_\odot$ we have $r = 5.6 \times 10^{10}$ cm, $p = 1.6 \times 10^{12}$ dyne/cm^2, and $\rho = 0.018$ g/cm^3, so $\mathcal{L}_0 \approx 2 \times 10^{41}$ erg/sec, as compared with the solar luminosity $L = 3.9 \times 10^{33}$ erg/sec. By a wide margin, this is a case of efficient convection. In general cases of efficient convection, Eq. (1.7.24) requires that $\nabla(r)$ is very close to the adiabatic value $\nabla_{\text{ad}}(r)$. In particular, where $\mathcal{E}$ is related to the pressure by Eq. (1.7.9), ∇_{ad} is given by Eq. (1.7.13), and so in the case of efficient convection we have

$$p(r) = K\rho^\Gamma(r), \tag{1.7.26}$$

where K is a constant that depends on conditions at the boundary of this region. This is the case throughout the convective region of the Sun, aside from a thin shell near the surface, where the pressure drops from 10^6 dyne/cm^2 to 10^5 dyne/cm^2. (But, as already mentioned, due to the effect of ionization, Γ is not constant in the outer parts of the convective region.) Where Eq. (1.7.26) holds throughout a star's interior, the star is known as a *polytrope*. Such stars are discussed further in Section 1.8.

There is another way of expressing this. The second law of thermodynamics tells us that there is a function s of ρ, p, etc. known as the *specific entropy*, or entropy per gram, for which[39]

$$T\,ds = d(\mathcal{E}/\rho) + p\,d(1/\rho). \tag{1.7.27}$$

Hence Eq. (1.7.8) can be interpreted as the statement that convection does not change the specific entropy of the convected fluid elements:

$$\delta s = 0. \tag{1.7.28}$$

---

[39] We are using δ to denote a change in a fluid element as it rises or falls in the star, while d stands for an arbitrary variation, not necessarily related to any actual motion.

This is because heat conduction is neglected here, which is generally a good approximation in stars. In regions where convection is efficient the specific entropy tends to a nearly uniform value to keep the convective energy transport consistent with the actual luminosity of the star. Stars with a uniform entropy per gram are said to be *isentropic*.

Though not strictly necessary for our purposes, it is instructive to work out a formula for the specific entropy for gases. With the internal energy given by Eq. (1.7.9), Eq. (1.7.27) reads

$$T\,ds = \frac{1}{\Gamma - 1}\left(\frac{dp}{\rho} + \Gamma p\,d\left(\frac{1}{\rho}\right)\right) = \frac{\rho^{\Gamma-1}}{\Gamma - 1} d\left(\frac{p}{\rho^{\Gamma}}\right). \tag{1.7.29}$$

For an ideal gas, $T = p/R\rho$, with R constant, so

$$ds = \frac{R}{\Gamma - 1}\left(\frac{\rho^{\Gamma}}{p}\right) d\left(\frac{p}{\rho^{\Gamma}}\right) = \frac{R}{\Gamma - 1} d\ln\left(\frac{p}{\rho^{\Gamma}}\right). \tag{1.7.30}$$

Hence

$$s = \frac{R}{\Gamma - 1} \ln\left(\frac{p}{\rho^{\Gamma}}\right) + \text{constant.} \tag{1.7.31}$$

We see again that p/ρ^{Γ} is constant in an isentropic star.

In typical stars there are regions stable against convection, in which energy transport is by radiation and p/ρ^{Γ} increases with r, and others with effective convection, in which p/ρ^{Γ} is constant. For instance, in the Sun there is a core with radiative energy transport, extending from the center where $p \simeq 2 \times 10^{17}$ dyne/cm^2, out to a radius about $0.65 R_{\odot}$ where the pressure has dropped to about 3×10^{13} dyne/cm^2. This is surrounded by an outer convective layer, and (since convection cannot carry energy into empty space) a relatively thin surface layer dominated by radiative energy transport. In more massive stars, there typically is a convective core, and an outer layer dominated by radiative transport that is stable against convection.

None of this affects the general results of Section 1.3 because the radii where regions of convective energy transport begin or end, and the values of p/ρ^{Γ} in these regions, are set by the conditions in the adjacent regions of radiative energy transport, and so are ultimately determined in terms of physical constants and the value of the nominal stellar radius R where the boundary conditions $\rho(R) = p(R) = 0$ are imposed. Also, the general results of Section 1.6 for the main sequence are unchanged, because nothing regarding convection involves new dimensionful constants.

* * * * *

For isentropic stars, whether or not satisfying the conditions for a polytrope, the equations of hydrostatic equilibrium can be expressed as a variational principle, which will prove useful when we come to stellar instability in Section 1.9. Let us consider the variation in the total energy

$$E = \int_0^R 4\pi r^2 \left(\mathcal{E}(r) - \frac{GM(r)\rho(r)}{r} \right) dr. \qquad (1.7.32)$$

Changes $\delta\rho$ and $\delta\mathcal{E}$ in the mass and energy densities produce a change in the total energy

$$\delta E = \int_0^R 4\pi r^2 \left(\delta\mathcal{E}(r) - \frac{GM(r)\,\delta\rho(r)}{r} - \frac{G\rho(r)}{r} \int_0^r 4\pi r'^2 \delta\rho(r')\,dr' \right) dr. \qquad (1.7.33)$$

In the first term, we use Eq. (1.7.8), which gives $\delta\mathcal{E} = (\mathcal{E} + p)\,\delta\rho/\rho$. In the third term, we interchange the order of integration, and also interchange the coordinate labels r and r'. This gives

$$\delta E = \int_0^R 4\pi r^2 \mathcal{F}(r)\,\delta\rho(r)\,dr, \qquad (1.7.34)$$

where

$$\mathcal{F}(r) = \frac{\mathcal{E}(r) + p(r)}{\rho(r)} - \frac{GM(r)}{r} - G\int_r^R 4\pi r'\rho(r')\,dr'. \qquad (1.7.35)$$

A straightforward calculation using Eq. (1.7.8) gives

$$\frac{d\mathcal{F}(r)}{dr} = \frac{1}{\rho(r)}\frac{dp(r)}{dr} + \frac{GM(r)}{r^2}. \qquad (1.7.36)$$

This vanishes according to the equation (1.1.4) of hydrostatic equilibrium, so $\mathcal{F}(r)$ is a constant $\mathcal{F}_0$, and therefore Eq. (1.7.32) reads

$$\delta E = \mathcal{F}_0 \int_0^R 4\pi r^2 \delta\rho(r)\,dr = \mathcal{F}_0\,\delta M. \qquad (1.7.37)$$

Thus, although the equation of hydrostatic equilibrium does not tell us that either E or M is stationary, it does tell us that E is stationary if M is. (The same result applies in general relativity,[40] with the total baryon number N_B times the baryon rest mass m_B taking the place of M and $M - m_B N_B$ taking the place of E.)

1.8 Polytropes

There are several classes of stars for which the pressure is simply proportional to a power of density, at least away from the surface:

$$p = K\rho^\Gamma, \qquad (1.8.1)$$

---

[40] For a textbook demonstration, see Section 11.2 of S. Weinberg, *Gravitation and Cosmology* (Wiley, New York, 1972).

with K and Γ constant throughout the star. Such stars are known as *polytropes* with index Γ. These include the following types,

- Ordinary stars with efficient convective energy transport. As shown in the previous section, these stars obey Eq. (1.8.1), with Γ typically close to 5/3, and K depending on boundary conditions, such as the values of the central density and pressure.
- As we shall see in Section 1.10, exceptionally light white dwarf stars obey Eq. (1.8.1) with Γ usually close to 5/3, and exceptionally heavy white dwarf stars obey Eq. (1.8.1) with $\Gamma \simeq 4/3$. In both cases K depends only on the chemical composition, as well as on fundamental physical constants.
- Supermassive stars. As discussed in Section 1.11, these stars obey Eq. (1.8.1) with $\Gamma \simeq 4/3$ and with K depending on the molecular weight and on the ratio of matter to radiation pressure, as well as on fundamental physical constants.

In this section we will treat all polytropes in common, not inquiring into the reason for Eq. (1.8.1).

Since the temperature does not enter in Eq. (1.8.1), we can work out the properties of the star using only the hydrostatic equations (1.1.4) and (1.1.5). It will be convenient now to rewrite these two first-order differential equations as a single second-order equation for the density:

$$\frac{d}{dr}\left(\frac{r^2}{\rho(r)}\frac{d}{dr}\rho^\Gamma(r)\right) + \frac{4\pi G}{K}r^2\rho(r) = 0. \tag{1.8.2}$$

As boundary conditions, we can take the central density to have some assumed value $\rho(0)$ and, since the analyticity of ρ as a function of $\mathbf{x}$ requires $\rho(r)$ to be a power series in r^2 near $r = 0$, we also take $\rho'(0) = 0$. With two initial conditions, we have a unique solution, depending only on Γ and on the free parameters K/G and $\rho(0)$.

There is an apparent paradox in the case of stars with efficient convective energy transport. Here there is not just one free stellar parameter, such as the star's mass or radius, but *two* free parameters, which can be taken as $\rho(0)$ and $K = p(0)/\rho(0)^\Gamma$. Thus the Vogt–Russell theorem mentioned in Section 1.3 does not apply to such polytropes. This may seem surprising, because we can think of the star as described by three first-order differential equations: Eqs. (1.1.4) and (1.1.5), together with

$$\frac{d}{dr}\left(\frac{p(r)}{\rho^\Gamma(r)}\right) = 0,$$

together with three parameter-free boundary conditions: $\mathcal{M}(0) = 0$, $\rho(R) = 0$, and $p(R) = 0$. So why, with an equal number of first-order differential equations and parameter-free boundary conditions, do we have any free parameters beyond the radius R at which some of the boundary conditions are imposed? The reason why this counting does not work here, though it may seem the same as the sort we used in Section 1.3, is that we are really imposing only one boundary condition at the surface. With $p(r)/\rho^\Gamma(r)$ constant, the condition $\rho(R) = 0$ *implies* that $p(R) = 0$. Having three first-order differential equations and only two independent parameter-free boundary conditions depending on R, there is an additional free parameter, which can be taken as K or $\rho(0)$, in addition to the radius R at which one of the boundary conditions is imposed.

Returning now to general polytropes, the free parameters in Eq. (1.8.2) can be eliminated by re-scaling the independent and dependent variables. First, define

$$\Theta \equiv \left(\frac{\rho(r)}{\rho(0)}\right)^{\Gamma-1}. \tag{1.8.3}$$

Then Eq. (1.8.2) gives

$$\frac{1}{r^2}\frac{d}{dr}\left(r^2\frac{d}{dr}\Theta\right) + \frac{4\pi G(\Gamma-1)}{K\Gamma}\rho(0)^{(2-\Gamma)}\Theta^{1/(\Gamma-1)} = 0.$$

We can get rid of the constant in the second term by introducing

$$\xi \equiv \left(\frac{4\pi G(\Gamma-1)}{K\Gamma}\right)^{1/2}\rho(0)^{(2-\Gamma)/2}r. \tag{1.8.4}$$

The differential equation (1.8.1) then becomes

$$\frac{1}{\xi^2}\frac{d}{d\xi}\left(\xi^2\frac{d}{d\xi}\Theta(\xi)\right) + \Theta(\xi)^{1/(\Gamma-1)} = 0, \tag{1.8.5}$$

and the boundary conditions are

$$\Theta(0) = 1, \quad \Theta'(0) = 0. \tag{1.8.6}$$

(The requirement $\Theta'(0) = 0$ like the requirement $\rho'(0) = 0$ is needed for the analyticity of $\rho(r)$ at $r = 0$ as a function of the Cartesian components of $\mathbf{x}$.)

Equation (1.8.5) is known as the *Lane–Emden equation*,[41] and was much studied in the early years of the twentieth century. It was shown that, for $\Gamma > 6/5$, its solution vanishes at a finite value ξ_1 of ξ, so the radius of the star is

$$R = \left(\frac{4\pi G(\Gamma-1)}{K\Gamma}\right)^{-1/2}\rho(0)^{-(2-\Gamma)/2}\xi_1. \tag{1.8.7}$$

The star's mass is

$$M = \int_0^R 4\pi r^2 \rho(r)\,dr$$

$$= 4\pi\rho(0)^{(3\Gamma-4)/2}\left(\frac{K\Gamma}{4\pi G(\Gamma-1)}\right)^{3/2}\int_0^{\xi_1}\xi^2\Theta^{1/(\Gamma-1)}(\xi)\,d\xi.$$

---

[41] The classic discussion of the Lane–Emden equation is by S. Chandrasekhar, *An Introduction to the Study of Stellar Structure* (University of Chicago Press, Chicago, IL, 1939).

By using Eq. (1.8.5), we easily see that

$$\int_0^{\xi_1} \xi^2 \Theta^{1/(\Gamma-1)}(\xi)\, d\xi = -\xi_1^2 \Theta'(\xi_1),$$

so

$$M = 4\pi\rho(0)^{(3\Gamma-4)/2} \left(\frac{K\Gamma}{4\pi G(\Gamma-1)}\right)^{3/2} \xi_1^2 |\Theta'(\xi_1)|. \tag{1.8.8}$$

There are just three values of $\Gamma > 1$ for which exact non-singular solutions of the Lane–Emden equation are known.

- For $\Gamma = \infty$, Eq. (1.8.5) is linear and inhomogeneous. The general solution is $-\xi^2/6$ plus any linear combination of $1/\xi$ and 1. The condition $\Theta(0) = 1$ fixes the solution to be simply $\Theta(\xi) = 1 - \xi^2/6$. This gives $\xi_1 = \sqrt{6}$ and $\xi_1^2 \Theta'(\xi_1) = -2\sqrt{6}$.
- For $\Gamma = 2$, Eq. (1.8.5) is linear and homogeneous. The general solution is any linear combination of $\sin\xi/\xi$ and $\cos\xi/\xi$. The condition $\Theta(0) = 1$ fixes the solution to be simply $\Theta(\xi) = \sin\xi/\xi$. This gives $\xi_1 = \pi$ and $\xi_1^2 \Theta'(\xi_1) = -\pi$.
- For $\Gamma = 6/5$, the solution of Eq. (1.8.5) with $\Theta(0) = 1$ is

$$\Theta(\xi) = (1 + \xi^2/3)^{-1/2}.$$

This reaches zero only at infinity, so $\xi_1 = \infty$, but $\xi^2 \Theta'(\xi)$ approaches the finite value $-\sqrt{3}$ for $\xi \to \infty$, so, though the radius is infinite, the mass is finite.

For other values of $\Gamma > 1$ a numerical computation is needed.[42] Here are some values of ξ_1 and $\xi_1^2 |\Theta'(\xi_1)|$ for several values of Γ:

Γ	ξ_1	$\xi_1^2 |\Theta'(\xi_1)|$
6/5	∞	$\sqrt{3}$
4/3	6.89685	2.01824
3/2	4.35287	2.41105
5/3	3.65375	2.71406
2	π	π
∞	$\sqrt{6}$	$2\sqrt{6}$

The isothermal case $\Gamma = 1$ is discussed in connection with galaxies in Section 4.2.

---

[42] Chandrasekhar, *op. cit.*

1.9 Instability

We noted in Section 1.1 that stars that are close to a polytrope with $\Gamma = 4/3$ are at the brink of a catastrophic instability. In this section we will prove a theorem that allows us to identify more precisely the threshold parameters at which such stars become unstable.

Suppose that a time-independent equilibrium stellar configuration is subject to an infinitesimal perturbation. As usual for perturbations of time-independent equilibrium, the perturbations $\delta\rho(\mathbf{x}, t)$, $\delta T(\mathbf{x}, t)$, etc. of various quantities can be expressed as a sum over normal modes, the contribution of each normal mode having a time-dependence given by a factor $e^{-i\omega t}$, with various values of ω (not necessarily real) for the various normal modes.[43] Each frequency ω is a function of the various parameters characterizing the equilibrium configuration, such as mass and/or central density.

In the absence of dissipative effects like heat conduction, the equations governing the time-dependence of the perturbations have the symmetry of time-reversal invariance, so that if $\delta\rho(\mathbf{x}, t)$, $\delta T(\mathbf{x}, t)$, etc. is a solution of these equations, then so is $\delta\rho(\mathbf{x}, -t)$, $\delta T(\mathbf{x}, -t)$, etc. This tells us that if ω is the frequency for some normal mode, then there is another normal mode with frequency $-\omega$. If ω is complex then $\exp(-i\omega t)$ grows exponentially unless the imaginary part of ω is negative, in which case $\exp(i\omega t)$ grows exponentially. Hence the equilibrium configuration is unstable unless all the frequencies ω characterizing the various normal modes are real.

Now, consider an equilibrium configuration with parameters for which all ω are real. Small perturbations will oscillate, but not grow. If we vary the star's parameters some ω may become complex, marking a transition to instability, but this faces an obstacle. Everything in these equations is real, so if $\delta\rho(\mathbf{x}, t)$, $\delta T(\mathbf{x}, t)$, etc. is a solution of these equations, then so is its complex conjugate $\delta\rho(\mathbf{x}, t)^*$, $\delta T(\mathbf{x}, t)^*$, etc., which tells us that if ω is the frequency for some normal mode, then there is another normal mode with frequency $-\omega^*$, as well as time-reversed modes with frequencies $-\omega$ and ω^*. Thus, if a generic real frequency ω became complex for some value of a stellar parameter, then the two modes with real frequencies ω and $-\omega$ would become four modes with frequencies ω, $-\omega$, $-\omega^*$, and ω^*. This is impossible; the number of modes is set by the dimensionality of the problem, and cannot suddenly increase or decrease.

---

[43] This is a consequence of the time-translation symmetry of the problem. If $\delta\rho(\mathbf{x}, t)$, $\delta T(\mathbf{x}, t)$, etc. is a solution of the differential equations for small perturbations, then so is $\delta\rho(\mathbf{x}, t + \delta t)$, $\delta T(\mathbf{x}, t + \delta t)$, etc. Since the equations governing these very small perturbations are linear, the solution at $t + \delta t$ must be a linear combination of the various solutions at t. By diagonalizing the matrix in this linear combination, we obtain an equal number of solutions in which each $\delta\rho(\mathbf{x}, t + \delta t)$, $\delta T(\mathbf{x}, t + \delta t)$, etc. is simply proportional to the corresponding $\delta\rho(\mathbf{x}, t)$, $\delta T(\mathbf{x}, t)$, etc., with a coefficient of proportionality that differs from unity by a term of first order in δt. That is, $\delta\rho(\mathbf{x}, t + \delta t) = [1 - i\omega\delta t]\delta\rho(\mathbf{x}, t)$, $\delta T(\mathbf{x}, t + \delta t) = [1 - i\omega\delta t]\delta T(\mathbf{x}, t)$, etc., with ω some constant, This implies the desired time-dependence, proportional to $\exp(-i\omega t)$.

There is, however, a way in which a real frequency ω can become complex, and the star thereby become unstable. If for some set of parameter values the two real frequencies ω and $-\omega$ come together, so that ω vanishes, then for slightly different parameters the frequency can become pure imaginary, so that $\omega = -\omega^*$, and there are still just two normal modes, with frequencies ω and $-\omega = \omega^*$. We conclude that the transition from stability, with all ω real, to instability, with some ω complex (actually imaginary), takes place for parameter values at which some ω vanishes.[44]

For the parameter values at which the ω for some normal mode vanishes, this normal mode becomes a time-independent perturbation of the stellar configuration, satisfying the equations of stellar structure. Since this perturbation becomes time-dependent for infinitesimal ω, it must preserve the values of conserved quantities, such as the total energy and baryon number. Thus (with the possible exceptions described in footnote 2) *a time-independent stellar configuration can become unstable only at values of stellar parameters at which there exists a time-independent perturbation that preserves the values of all conserved parameters.*

In cases where the effects of general relativity can be neglected at the transition to instability, we can take the two quantities that have to be conserved as the energy E, not counting rest masses, and the total rest mass M, defined equal to the baryon number B times the rest mass m_B per baryon. As we saw at the end of Section 1.7, at least for stars with a uniform entropy per rest mass, if one of these is stationary the other is too, so we can concentrate on perturbations that leave just E conserved. We will see an example of this in Section 1.10 for iron white dwarfs.

There are other cases, where the instability arises because of effects of general relativity. Here again there are two conserved quantities, the mass M in the Schwarzschild metric (see Eq. (1.9.A1) below) and the total baryon number B. As in the non-relativistic case, it is especially convenient to look for values of stellar parameters at which the total internal energy $E \equiv M - m_B c^2 B$ (which includes gravitational energy and everything else except the energy in rest masses) is stationary. Obviously if M and B are stationary then so is E, and the theorem mentioned at the end of Section 1.7 tells us that at least for stars with a uniform entropy per baryon, in general relativity the condition that E is stationary is sufficient as well as necessary for both M and B to be stationary.

---

[44] Strictly speaking, there are other possible ways in which, at the transition to instability, several modes may come together to have the same frequency. As an example, suppose that for some set of parameters we have four normal modes with distinct real frequencies ω_1, $-\omega_1$, ω_2, and $-\omega_2$. If we vary the parameters in such a way that ω_1 and ω_2 become equal to the same real value ω_0, then for a further variation of parameters we could again have four distinct frequencies, $\omega_0 + i\epsilon$, $-\omega_0 - i\epsilon$, $-\omega_0 + i\epsilon$, and $\omega_0 - i\epsilon$ with $\epsilon \neq 0$ real. For instance, this happens if the frequencies of the four normal modes are the roots of the equation $(\omega^2 - a^2)^2 - b = 0$, with a and b real. The roots are real for $b > 0$, but become complex (though not pure imaginary) as b moves to negative values. I am not aware of these possibilities actually occurring in stars, and they will not be considered in what follows.

To find where E is stationary in a relativistic context, we will use an expansion for E in powers of the dimensionless quantities $p/\rho c^2$ and $G\mathcal{M}/rc^2$. These two quantities according to Eq. (1.1.4) are roughly of the same order of magnitude, which will be denoted v^2/c^2, and are assumed to be very small.[45] The expansion reads

$$E = \int_0^R \mathcal{E}(r)\, 4\pi r^2\, dr - 3\int_0^R p(r)\, 4\pi r^2\, dr$$

$$+ \int_0^R 6\pi r^4 dr\, p'^2(r)/c^2 \rho(r)$$

$$- \int_0^R 8\pi r^3 dr\, p(r) p'(r)/c^2 \rho(r) + \cdots, \qquad (1.9.1)$$

where $\mathcal{E}$ is the thermal energy density, excluding only gravitational energy and the energy in rest masses. The individual terms on the first line are of order $\mathcal{M}v^2$, while the terms on the second and third lines are of order $\mathcal{M}v^4/c^2$, and the dots denote terms no larger than of order $\mathcal{M}v^6/c^4$.

The derivation of the expansion (1.9.1) is given at the end of this section. This derivation does not rely on any assumption about the star being a polytrope, but the expansion finds its most important application when the terms of order $\mathcal{M}v^2$ on the first line nearly cancel, so that the relativistic corrections on the second and third lines become important. This occurs when the pressure is close to $\mathcal{E}/3$, i.e., when the star is close to a polytrope with $\Gamma = 4/3$. As already remarked in Section 1.1, when a star is very close to having $\mathcal{E} = 3p$, as for a polytrope with $\Gamma \simeq 4/3$, very small corrections to a stellar model can make the difference between stability and instability. Because the relativistic corrections on the second and third lines of Eq. (1.9.1) are already much smaller than the individual terms on the first line, in the case at hand these terms can be calculated using the non-relativistic equations of stellar structure for a polytrope with $\Gamma = 4/3$, that is with $p = K\rho^{4/3}$ for some constant K, the inaccuracies in this calculation being even smaller. Using Eqs. (1.8.3), (1.8.4), and (1.8.1), we easily find that Eq. (1.9.1) becomes

$$E = \int_0^R 4\pi r^2\, dr \big[\mathcal{E}(r) - 3p(r)\big] + \frac{16\pi}{(\pi G)^{3/2} c^2}\rho^{2/3}(0) K^{7/2} \eta, \qquad (1.9.2)$$

---

[45] For ordinary stars supported by non-relativistic gas pressure, $p/\rho c^2 \approx v_{\text{th}}^2/c^2$, where v_{th} is a typical thermal velocity, generally much less than c. Even when the pressure is dominated by relativistic particles, such as electrons in the most massive white dwarfs or photons in very massive stars, the ratio $p/\rho c^2$ is of the order of the ratio of the energy in these relativistic particles to the energy in baryon rest masses, and is still generally very small.

where η is the positive numerical constant[46]

$$\eta = -2\int_0^{\xi_1} \Theta^4(\xi)\Theta'(\xi)\xi^3\,d\xi + 6\int_0^{\xi_1} \Theta^3(\xi)\Theta'^2(\xi)\xi^4\,d\xi = 3.49815, \quad (1.9.3)$$

with $\xi_1 = 6.89685$ corresponding to the radius at the star's surface. We still need to calculate the first term in Eq. (1.9.2) separately for individual cases, such as massive white dwarfs and supermassive stars, as will be done in Sections 1.10 and 1.11, but the second term in Eq. (1.9.2) represents a universal relativistic correction for stars that are close to polytropes with $\Gamma = 4/3$.

Appendix: Derivation of Relativistic Correction to Energy

As in the non-relativistic case, in general relativity there are two quantities that must be conserved, at least in any spherically symmetric perturbations of the star, and that therefore must both be stationary at values of parameters such as central density at which there is a transition from stability to instability. One of them is the mass M appearing in the Schwarzschild solution for the metric outside the star:

$$-g_{tt} = g_{rr}^{-1} = 1 - 2MG/rc^2, \quad g_{\theta\theta} = r^2, \quad g_{\phi\phi} = r^2\sin^2\theta. \quad (1.9.\text{A}1)$$

(Here and below we are using "standard" coordinates, for which inside or outside the star g_{tt} and g_{rr} are functions of r and t, while $g_{\phi\phi}$ and $g_{\theta\theta}$ are the same as for a flat space.) The Schwarzschild solution gives

$$M = \int_0^R \rho(r)\,4\pi r^2\,dr, \quad (1.9.\text{A}2)$$

where now $\rho(r)c^2$ is the total energy density (that is, the time–time component of the energy-momentum tensor $T^{\mu\nu}$), including mass energy and everything else except gravitational energy. (Gravitational energy is included in M in the difference between $4\pi r^2\,dr$ and the spatial volume element $4\pi r^2\sqrt{g_{rr}}\,dr$.)

The other conserved quantity is B, the total baryon number of the star:

$$B = \int_0^R B^t(r)\,4\pi r^2\sqrt{-g_{rr}(r)g_{tt}(r)}\,dr, \quad (1.9.\text{A}3)$$

where B^μ is the conserved current of baryon number. We can write B^t in terms of the scalar baryon density $n \equiv U_\mu B^\mu$, where U^μ is the velocity four-vector, normalized so that $U^\mu U^\nu g_{\mu\nu} = -1$. For a fluid at rest $U^r = U^\theta = U^\phi = 0$, so $U^t = 1/\sqrt{-g_{tt}}$, $U_t = \sqrt{-g_{tt}}$, $n = \sqrt{-g_{tt}}B^t$, and therefore

---

[46] The numerical value given here is inferred from Eqs. (6.9.29)–(6.9.31) of S. Shapiro and S. Teukolsky, *Black Holes, White Dwarfs, and Neutron Stars* (Wiley, New York, 1983).

$$B = \int_0^R n(r)\, 4\pi r^2 \sqrt{g_{rr}(r)}\, dr. \tag{1.9.A4}$$

The Schwarzschild solution inside the star gives

$$g_{rr}(r) = \left(1 - \frac{2GM(r)}{rc^2}\right)^{-1}, \tag{1.9.A5}$$

where, as before,

$$M(r) = \int_0^r 4\pi r'^2 \rho(r')\, dr'. \tag{1.9.A6}$$

Thus

$$B = \int_0^R n(r)\, 4\pi r^2 \left(1 - \frac{2GM(r)}{rc^2}\right)^{-1/2} dr. \tag{1.9.A7}$$

Of course, in the non-relativistic limit ρ is the rest mass density, and $M = m_B B$, where $m_B \simeq 938$ MeV/c^2 is the rest mass per baryon. More generally, we can define an internal energy density $\mathcal{E}$ excluding rest masses, by

$$\rho(r) \equiv n(r) m_B + \mathcal{E}(r)/c^2. \tag{1.9.A8}$$

We will eventually be assuming that the star, though not necessarily a polytrope, is close to a non-relativistic polytrope with $\Gamma = 4/3$, and therefore has a pressure $p(r)$ close to $3\mathcal{E}(r)$. Without yet making this approximation, we can anticipate that it will be convenient to express ρ as

$$\rho(r)c^2 = m_B n(r) c^2 + 3p(r) + \Delta\mathcal{E}(r), \tag{1.9.A9}$$

where

$$\Delta\mathcal{E}(r) \equiv \mathcal{E}(r) - 3p(r), \tag{1.9.A10}$$

which will eventually be treated as a small perturbation, arising from the finite electron mass for white dwarfs and the finite baryon kinetic energy for supermassive stars.

It is important to be clear about the order of magnitude of the terms in M and $m_B B$. The leading term in both, $\int_0^R 4\pi r^2 m_B n(r)\, dr$, is the non-relativistic approximation to M, and hence is of order M. The next-to-leading terms, $\int_0^R 12\pi r^2 p(r)\, dr/c^2$ in M and $\int_0^R 4\pi r^2 m_B n(r)\, (GM(r)/c^2 r)\, dr$ in $m_B B$, are of order Mv^2/c^2, where v is a characteristic gas particle velocity, with $v^2 \approx GM/r \approx p/m_B n$, assumed much less than c. The general relativistic correction $\int_0^R 4\pi r m_B n(r)\, (GM(r)/rc^2)^2\, dr$ in $m_B B$ is of order Mv^4/c^4.

To eliminate the terms of order M, we consider the difference, which gives the total internal energy E:

$$E/c^2 \equiv M - m_B B$$
$$= \int_0^R 4\pi r^2 \, dr \left[3p(r)/c^2 + \Delta\mathcal{E}(r)/c^2 \right.$$
$$\left. + m_B n(r)\left(1 - (1 - 2GM(r)/rc^2)^{-1/2}\right) \right]. \quad (1.9.\text{A}11)$$

The terms of order Mv^2/c^2 also cancel in Eq. (1.9.A11), as can be seen by integrating the pressure term by parts,

$$\int_0^R 12\pi r^2 p(r) \, dr = \int_0^R p(r) \, d(4\pi r^3) = -\int_0^R p'(r) \, 4\pi r^3 \, dr,$$

and then using the relativistic equilibrium condition[47]

$$-r^2 p'(r) = G\big(\rho(r) + p(r)/c^2\big)\big(\mathcal{M}(r) + 4\pi r^3 p(r)/c^2\big)\big(1 - 2\mathcal{M}(r)G/rc^2\big)^{-1}, \quad (1.9.\text{A}12)$$

which together with formula (1.9.A9) for the total energy density gives

$$\int_0^R 12\pi r^2 p(r) \, dr = \int_0^R 4\pi r G\big(m_B n(r) + 4p(r)/c^2 + \Delta\mathcal{E}(r)/c^2\big)$$
$$\times \big(\mathcal{M}(r) + 4\pi r^3 p(r)/c^2\big)\big(1 - 2\mathcal{M}(r)G/rc^2\big)^{-1} dr. \quad (1.9.\text{A}13)$$

Using this for the first term in Eq. (1.9.A11) gives the internal energy

$$E = \int_0^R 4\pi r^2 \left[\Delta\mathcal{E}(r) + m_B n(r)c^2\big(1 - (1 - 2GM(r)/rc^2)^{-1/2}\big) \right] dr$$
$$+ \int_0^R 4\pi r G\big(m_B n(r)c^2 + 4p(r) + \Delta\mathcal{E}(r)\big)$$
$$\times \big(\mathcal{M}(r) + 4\pi r^3 p(r)/c^2\big)\big(1 - 2\mathcal{M}(r)G/rc^2\big)^{-1} dr. \quad (1.9.\text{A}14)$$

So far, although we have been guided by order-of-magnitude estimates, the result (1.9.A14) is exact. Now, we note the term of order Mv^2 in the first line is $-\int_0^R 4\pi r^2 m_B n (GM/rc^2) \, dr$ and cancels the term of order Mv^2/c^2 in the second line, which is $+\int_0^R 4\pi r G m_B n \mathcal{M} \, dr/c^2$. The leading terms are then of order Mv^4/c^2:

---

[47] For a textbook derivation, see Section 11.1 of S. Weinberg, *Gravitation and Cosmology* (Wiley, New York, 1972).

$$E \simeq \int_0^R 4\pi r^2 \left[\Delta\mathcal{E}(r) - \frac{3}{2} m_B n(r) c^2 (GM(r)/rc^2)^2 \right] dr$$
$$+ G \int_0^R 4\pi r \, dr \Big[m_B n(r) M(r) (2M(r)G/rc^2)$$
$$+ m_B n(r) 4\pi r^3 p(r) + 4p(r) M(r) \Big]. \quad (1.9.\text{A}15)$$

Because each relativistic correction term in Eq. (1.9.A15) is individually small, of order Mv^4/c^4, they can each be evaluated by using the non-relativistic approximation

$$\rho(r) = m_B n(r) \quad (1.9.\text{A}16)$$

and the non-relativistic equation of equilibrium, Eq. (1.1.4), which gives

$$M(r) = -r^2 p'(r)/G\rho(r). \quad (1.9.\text{A}17)$$

Then, also combining the last term on the first line of Eq. (1.9.A15) with the first term on the second line,

$$E = \int_0^R 4\pi r^2 \, dr \, \Delta\mathcal{E}(r) + \int_0^R 2\pi r^4 \, dr \, p'^2(r)/c^2 \rho(r)$$
$$+ \int_0^R 16\pi^2 G r^4 \, dr \, \rho(r) p(r)/c^2 - \int_0^R 16\pi r^3 \, dr \, p(r) p'(r)/c^2 \rho(r).$$
$$(1.9.\text{A}18)$$

This can be further simplified, by noting that to order Mv^4/c^2 the third term is a linear combination of the second and fourth terms. Using the non-relativistic equation (1.1.4) of hydrostatic equilibrium (which is justified since this term is already small), we have

$$4\pi G r^2 \rho = G \frac{dM}{dr} = -\frac{d}{dr}\left(\frac{r^2 \, dp}{\rho \, dr}\right),$$

so integrating by parts gives

$$\int_0^R 16\pi^2 G r^4 \, dr \, \rho(r) p(r) = -\int_0^R 4\pi r^2 p(r) \frac{d}{dr}\left(\frac{r^2}{\rho(r)} p'(r)\right)$$
$$= \int_0^R 4\pi r^4 p'(r)^2/\rho(r)$$
$$+ \int_0^R 8\pi r^3 p(r) p'(r)/\rho(r), \quad (1.9.\text{A}19)$$

so that (1.9.A18) becomes

$$E = \int_0^R 4\pi r^2 \, dr \, \Delta \mathcal{E}(r) + \int_0^R 6\pi r^4 \, dr \, p'^2(r)/c^2 \rho(r)$$
$$- \int_0^R 8\pi r^3 \, dr \, p(r) p'(r)/c^2 \rho(r), \qquad (1.9.\text{A}20)$$

as was to be shown.

1.10 White Dwarfs and Neutron Stars

In a white dwarf star nuclear reactions have come to an end, the star has cooled to the point that temperature may be neglected in studying the interior, and pressure and kinetic energy are provided by cold degenerate electrons. To a good approximation, the mass density ρ is $m_1 \mu$ times the electron number density, where $m_1 = 931.49$ GeV/c^2 is the nuclear mass for unit atomic weight, and here $\mu \equiv A/Z$ is the atomic weight per electron, equal to 55.847/26 for iron. According to the rules of Fermi statistics, this gives the mass density as[48]

$$\rho(r) = \frac{8\pi m_1 \mu}{h^3} \int_0^{k_\text{F}(r)} k^2 \, dk = \frac{8\pi m_1 \mu k_\text{F}^3(r)}{3h^3}, \qquad (1.10.1)$$

where k_F is the maximum momentum of the filled electron levels, known as the Fermi momentum, and $h = 2\pi \hbar$ is the original Planck constant. (The extra factor 2 in 8π takes account of the electron's two spin states.) The internal energy density (excluding rest masses) and pressure of the electrons are then

$$\mathcal{E}(r) = \frac{8\pi}{h^3} \int_0^{k_\text{F}(r)} \left[\sqrt{k^2 c^2 + m_\text{e}^2 c^4} - m_\text{e} c^2 \right] k^2 \, dk \qquad (1.10.2)$$

and

$$p(r) = \frac{8\pi c^2}{3h^3} \int_0^{k_\text{F}(r)} \frac{k^4}{\sqrt{k^2 c^2 + m_\text{e}^2 c^4}} \, dk. \qquad (1.10.3)$$

Using Eq. (1.10.1) to express k_F in terms of ρ/μ, Eqs. (1.10.2) and (1.10.3) become formulas for $\mathcal{E}$ and p in terms of ρ/μ (or each other). With μ assumed uniform in the star, for any given μ and central density $\rho(0)$, we can use the

---

[48] Fermi statistics requires that no two electrons can have the same momentum and spin. The possible states of free particles are represented by wave functions of the form $\exp(i\mathbf{k} \cdot \mathbf{x}/\hbar)$, with $\mathbf{k}$ the momentum. To confine these particles in a finite volume L^3 without violating translation invariance, we require the wave function to be the same on opposite faces of a box with edge L, so that $\mathbf{k} = 2\pi \hbar \mathbf{n}/L$, where $\mathbf{n}$ is a vector with integer components. The number of such vectors $\mathbf{n}$ with magnitude between n and $n+dn$ is $4\pi n^2 \, dn$, so the number of possible momenta with magnitude between k and $k + dk$ is $(2\pi\hbar/L)^{-3} \times 4\pi k^2 \, dk$, and with two particles per momentum state, the number of particles per volume with momenta between k and $k + dk$ is $8\pi k^2 \, dk/(2\pi\hbar)^3$.

1.10 White Dwarfs and Neutron Stars

equations (1.1.4) and (1.1.5) of hydrostatic equilibrium to find $\rho(r)$ and $p(r)$ throughout the star, and in particular to find the stellar mass M and radius R.

In general a white dwarf star is not a polytrope, with $\mathcal{E}$ proportional to p and p proportional to a power of ρ, except in the limit of very small or very large density and Fermi momentum. According to Eq. (1.10.1), the critical density at which the Fermi momentum becomes equal to $m_e c$ is

$$\rho_c = \frac{8\pi m_1 \mu m_e^3 c^3}{3h^3} = 0.97 \times 10^6 \, \mu \text{ g/cm}^3. \tag{1.10.4}$$

For $\rho \ll \rho_c$ we have $k_F \ll m_e c$, so Eqs. (1.10.2) and (1.10.3) give

$$p = \frac{8\pi}{3m_e h^3} \int_0^{k_F} k^4 \, dk = \frac{8\pi k_F^5}{15 m_e h^3} = \frac{8\pi}{15 m_e h^3} \left(\frac{3h^3 \rho}{8\pi m_1 \mu} \right)^{5/3} \tag{1.10.5}$$

and $\mathcal{E} = 3p/2$. This is a polytrope, with $\Gamma = 5/3$, and

$$K = \frac{8\pi}{15 m_e h^3} \left(\frac{3h^3}{8\pi m_1 \mu} \right)^{5/3}. \tag{1.10.6}$$

Equations (1.8.7) and (1.8.8) then give the radius and mass of the star as

$$R = 3.65375 \times \left(\frac{8\pi G}{5K} \right)^{-1/2} \rho(0)^{-1/6} = 2.0 \times 10^4 \, \mu^{-1} \left(\frac{\rho(0)}{\rho_c} \right)^{-1/6} \text{ km} \tag{1.10.7}$$

and

$$M = 2.71406 \times 4\pi \rho(0)^{1/2} \left(\frac{5K}{8\pi G} \right)^{3/2} = 2.79 \mu^{-2} \left(\frac{\rho(0)}{\rho_c} \right)^{1/2} M_\odot. \tag{1.10.8}$$

Thus low-mass white dwarfs, with $\rho(0) \ll \rho_c$, have radii somewhat greater than the Earth's, and masses somewhat less than the Sun's. Also, their thermal plus gravitational energy (1.1.13) is

$$E = \int_0^R (\mathcal{E}(r) - 3p(r)) 4\pi r^2 \, dr = -6\pi \int_0^R p(r) r^2 \, dr$$

$$= -6\pi \left(\frac{5}{8\pi G} \right)^{3/2} K^{-1/2} \rho(0)^{7/6} \int_0^{\xi_1} \Theta^{5/2}(\xi) \xi^2 \, d\xi, \tag{1.10.9}$$

where $\xi_1 = 3.65375$.

For $\rho \gg \rho_c$ we have $k_F \gg m_e c$, so Eqs. (1.10.2) and (1.10.3) give

$$p = \frac{8\pi c}{3h^3} \int_0^{k_F} k^3 \, dk = \frac{2\pi c k_F^4}{3h^3} = \frac{2\pi c}{3h^3} \left(\frac{3h^3 \rho}{8\pi m_1 \mu} \right)^{4/3} \tag{1.10.10}$$

and $\mathcal{E} = 3p$. This is a polytrope, with $\Gamma = 4/3$, and

$$K = \frac{2\pi c}{3h^3}\left(\frac{3h^3}{8\pi m_1 \mu}\right)^{4/3}. \qquad (1.10.11)$$

Equations (1.8.7) and (1.8.8) then give the radius and mass of the star as

$$R = 6.89685 \times \left(\frac{\pi G}{K}\right)^{-1/2} \rho(0)^{-1/3} = 5.3 \times 10^4 \mu^{-1}\left(\frac{\rho(0)}{\rho_c}\right)^{-1/3} \text{ km} \qquad (1.10.12)$$

and

$$M = 2.01824 \times 4\pi \left(\frac{K}{\pi G}\right)^{3/2} = 5.87 \mu^{-2} M_\odot. \qquad (1.10.13)$$

It is striking that although R decreases and M increases with increasing central density, the mass approaches the limiting value (1.10.13), known as the *Chandrasekhar bound*. Of course, with $\Gamma = 4/3$, the energy (1.1.13) is $E = 0$.

White dwarfs with $\rho(0) \ll \rho_c$ have Γ considerably above $4/3$, so according to the arguments in Section 1.1 they are stable at least against complete dispersal. Also, Eqs. (1.10.8) and (1.10.9) show that in this region M and E both vary monotonically with $\rho(0)$, while according to the theorem cited in the previous section, in order for a white dwarf to become unstable with increasing central density it is necessary for $\rho(0)$ to reach a value at which the conserved quantities M and E have vanishing derivatives with respect to central density.

If our results so far were the whole story, white dwarfs would also be stable for $\rho(0) \gg \rho_c$. For $\Gamma = 4/3$ both E and M are constants, but as we shall see, by itself the small departure from the polytropic equation of state due to the finite electron mass would give both $-E$ and M a continued monotonic increase with $\rho(0)$. But there are two complications that make instability possible.

One complication is provided by neutronization: the baryon number per electron μ is not really constant. For sufficiently large central density, the Fermi momentum is large enough for it to be energetically favorable for electrons to be absorbed by protons, in the reaction $e^- + p \to \nu_e + n$. For an iron white dwarf, this occurs when $\rho(0)$ exceeds 1.14×10^9 g/cm^3, and has the effect that ^{56}Fe nuclei with $\mu = 2.15$ are converted to ^{56}Mn, with $\mu = 2.24$. The increase in μ eventually causes the mass (1.10.13) to stop rising toward a limit, and instead to reach a maximum close to the value (1.10.13), and then begin to decrease. This maximum marks the transition to instability, and thus represents the true upper bound on the masses of iron white dwarfs.

Another complication arises from general relativity. According to the results of the previous section, even if μ is constant there is a transition to instability at a value of central density for which a maximum is reached by the energy

$$E = \int 4\pi r^2 \, dr \left[\mathcal{E}(r) - 3p(r)\right] + \frac{16\pi}{(\pi G)^{3/2} c^2} \rho^{2/3}(0) K^{7/2} \eta, \qquad (1.10.14)$$

1.10 White Dwarfs and Neutron Stars

where $\eta = 3.49815$. The second term is a universal general relativistic correction for stars with $\mathcal{E}$ near $3p$, except that we must use the appropriate value for K, which for white dwarfs is given by Eq. (1.10.11). To calculate the first term for white dwarfs with $\rho(0) \gg \rho_c$, we use Eqs. (1.10.2) and (1.10.3) for $m_e^2 c^2 \ll k_F^2$, together with the familiar expansions

$$\sqrt{k^2 c^2 + m_e^2 c^4} = kc + m_e^2 c^3/2k + \cdots,$$

$$1/\sqrt{k^2 c^2 + m_e^2 c^4} = 1/kc - m_e^2 c/2k^3 + \cdots.$$

Taking k_F from Eq. (1.10.1), we obtain expansions in powers of electron mass[49]

$$\mathcal{E} = \frac{3hc}{4}\left(\frac{3}{8\pi}\right)^{1/3}\left(\frac{\rho}{\mu m_1}\right)^{4/3} - \left(\frac{m_e}{\mu m_1}\right)\rho c^2$$
$$+ \frac{3m_e^2 c^3}{4h}\left(\frac{3}{8\pi}\right)^{-1/3}\left(\frac{\rho}{\mu m_1}\right)^{2/3} + \cdots, \qquad (1.10.15)$$

$$3p = \frac{3hc}{4}\left(\frac{3}{8\pi}\right)^{1/3}\left(\frac{\rho}{\mu m_1}\right)^{4/3} - \frac{3m_e^2 c^3}{4h}\left(\frac{3}{8\pi}\right)^{-1/3}\left(\frac{\rho}{\mu m_1}\right)^{2/3} + \cdots. \qquad (1.10.16)$$

The leading terms give the equation of state (1.10.10) of a $\Gamma = 4/3$ polytrope. The terms in Eqs. (1.10.15) and (1.10.16) of first and second order in m_e give

$$\Delta \mathcal{E} \equiv \mathcal{E} - 3p = -\left(\frac{m_e}{\mu m_1}\right)\rho c^2 + \frac{3m_e^2 c^3}{2h}\left(\frac{3}{8\pi}\right)^{-1/3}\left(\frac{\rho}{m_1 \mu}\right)^{2/3} + \cdots$$

$$= -\left(\frac{m_e}{\mu m_1}\right)\rho c^2 + \frac{3m_e^2 c^4}{8m_1^2 \mu^2 K}\rho^{2/3} + \cdots. \qquad (1.10.17)$$

Since the factor m_e^2/m_1^2 makes the term in $\int 4\pi r^2 \, dr \, \Delta \mathcal{E}$ of second order in m_e very small, it can be evaluated by using the solution given in Section 1.8 for a non-relativistic polytrope with $\Gamma = 4/3$. Equation (1.10.17) then gives the expansion

$$\int 4\pi r^2 \, dr \, \Delta \mathcal{E} = -\left(\frac{m_e}{m_1 \mu}\right) Mc^2 + \frac{3\pi m_e^2 c^4 K^{1/2} \zeta}{2m_1^2 \mu^2 (\pi G)^{3/2}}\rho(0)^{-1/3} + \cdots, \qquad (1.10.18)$$

where[50]

$$\zeta = \int_0^{\xi_1} \Theta^2(\xi)\xi^2 \, d\xi = 4.3267. \qquad (1.10.19)$$

---

[49] The term in Eq. (1.8.15) of first order in m_e is present because we have chosen to define E excluding all rest masses, including the electron rest mass. As we shall see in Eq. (1.10.18), it leads to a term in E that is independent of central density, and therefore has no effect on the threshold for instability.

[50] The numerical value here is inferred from Eqs. (6.10.19) and (6.10.20) of S. Shapiro and S. Teukolsky, *Black Holes, White Dwarfs, and Neuton Stars* (Wiley, New York, 1983).

The critical central density ρ_{inst} for a transition from stability to instability is then the stationary point of Eq. (1.10.14):

$$0 = \left.\frac{\partial E}{\partial \rho(0)}\right|_{\rho(0)=\rho_{\text{inst}}} = -\frac{\pi m_e^2 c^4 K^{1/2} \zeta}{2m_1^2 \mu^2 (\pi G)^{3/2}} \rho_{\text{inst}}^{-4/3} + \frac{32\pi}{3(\pi G)^{3/2} c^2} \rho_{\text{inst}}^{-1/3} K^{7/2} \eta. \tag{1.10.20}$$

This gives[51]

$$\rho_{\text{inst}} = \frac{3m_e^2 c^6 \zeta}{64 m_1^2 \mu^2 K^3 \eta} = \frac{8\pi m_e^2 m_1^2 c^3 \mu^2 \zeta}{h^3 \eta} = 6.6 \times 10^9 \mu^2 \text{ g/cm}^3. \tag{1.10.21}$$

The critical densities for general relativistic instability along with neutronization thresholds[52] are given (both in g/cm^3) for three commonly considered chemical compositions in the table below. Since white dwarfs of low central density are stable, the transition to instability for increasing central density occurs at the *lower* of the neutronization threshold and ρ_{inst}. This transition is evidently produced by neutronization for ^{56}Fe, and by general relativity for ^{12}C and ^{4}He.

Critical densities and neutronization thresholds

Composition	Neutronization threshold	ρ_{inst}
^{56}Fe	1.14×10^9	3.06×10^{10}
^{12}C	3.9×10^{10}	2.63×10^{10}
^{4}He	1.37×10^{14}	2.63×10^{10}

Neutronization in a white dwarf star can only go so far before the star becomes unstable. But when a star that is too massive to form a stable white dwarf exhausts its nuclear fuel it collapses, becoming a supernova. The density increases, and the rapid rise in the electron Fermi momentum forces a nearly complete neutronization. Almost all of the star's protons and electrons are converted to neutrons, with just enough electrons left for the neutron decay $n \to p + e^- + \nu$ to be blocked by the Pauli exclusion principle, and with an equal number of protons left over to balance the electron charges. After blowing off enough matter, what remains is a stable neutron star.[53]

In a neutron star, it is neutrons rather than electrons that fill all quantum levels up to a Fermi momentum $k_F(r)$. Here the mass density is given again by Eq. (1.10.1), but with the neutron mass m_n in place of $m_1 \mu$:

---

[51] When cancellations and different notation are taken into account, this formula turns out to be identical to the second line of Eq. (6.10.28) of Shapiro and Teukolsky, *op. cit.*, derived in a rather different manner. They give a numerical result $6.615 \times 10^9 \mu^2$ g/cm^3 for ρ_{inst}.
[52] Thresholds are taken from Shapiro and Teukolsky, *op. cit.*, Table 3.1.
[53] W. Baade and F. Zwicky, *Phys. Rev.* **46**, 76 (1934).

1.10 White Dwarfs and Neutron Stars

$$\rho(r) = \frac{8\pi m_n}{h^3} \int_0^{k_F(r)} k^2 \, dk = \frac{8\pi m_n k_F^3(r)}{3h^3}. \tag{1.10.22}$$

There is again a critical density here, but now one for which the Fermi momentum becomes $m_n c$ rather than $m_e c$:

$$\rho_c = \frac{8\pi m_n^4 c^3}{3h^3} = 6.11 \times 10^{15} \text{ g/cm}^3. \tag{1.10.23}$$

The mean separation between neutrons is

$$\left(\frac{\rho}{m_n}\right)^{-1/3} = \left(\frac{\rho_c}{\rho}\right)^{1/3} \times 0.52 \times 10^{-13} \text{ cm}, \tag{1.10.24}$$

which for $\rho \ll \rho_c$ is greater than the range of nuclear forces, justifying the treatment of neutrons as free particles, as implicitly assumed in Eq. (1.10.22). Also, for $\rho \ll \rho_c$ even neutrons at the top of the Fermi sea are moving non-relativistically, so the neutron pressure is given by the same formula (1.10.5) as for low-mass white dwarfs, but with m_n in place of both m_e and $m_1 \mu$:

$$p = \frac{8\pi}{15 m_n h^3} \left(\frac{3h^3 \rho}{8\pi m_n}\right)^{5/3}. \tag{1.10.25}$$

This is again a polytrope with $\Gamma = 5/3$. Since the neutrons are moving non-relativistically, the structure of the neutron star for $\rho(0) \ll \rho_c$ is governed by the Newtonian equations of gravitation and dynamics, just like the structure of white dwarfs, and can therefore be treated by the methods of Section 1.8. In particular, we can use Eqs. (1.10.6)–(1.10.8) for the neutron star's radius R and mass M, again with m_n in place of both m_e and $m_1 \mu$:

$$R = 3.65375 \times (3Gm_n h^3)^{-1/2} \left(\frac{3h^3}{8\pi m_n}\right)^{5/6} \rho(0)^{-1/6} = 11.0 \text{ km} \times \left(\frac{\rho_c}{\rho(0)}\right)^{1/6} \tag{1.10.26}$$

and

$$M = 2.71406 \times 4\pi \rho(0)^{1/2} (3Gm_n h^3)^{-3/2} \left(\frac{3h^3}{8\pi m_n}\right)^{5/2} = 2.7 M_\odot \times \left(\frac{\rho(0)}{\rho_c}\right)^{1/2}. \tag{1.10.27}$$

The mass is again a few solar masses, like a white dwarf, but now in a radius of a few kilometers instead of a few thousand kilometers.

It may be surprising that both white dwarfs and neutron stars typically have masses of order $M_\odot$, even though they are supported by the degeneracy pressure of particles of very different mass: electrons for white dwarfs, and neutrons for neutron stars. This is because the electron mass cancels in Eq. (1.10.8) for white dwarf masses (though not in Eq. (1.10.7) for white dwarf radii). Indeed, Eqs. (1.10.8) and (1.10.27) give the masses of both white dwarfs and neutron stars as equal to different factors of order unity times the same combination of fundamental constants:

$$(\hbar c/G)^{3/2} m_1^{-3/2} = 1.90 M_\odot.$$

In the following section we will see the same combination of constants appearing in the mass of stars supported by radiation pressure.

The theory of neutron stars is much more complicated for $\rho(0)$ comparable to or greater than ρ_c. Here the neutron velocities are comparable to c, and since the neutrons are the source of the star's gravitational field, general relativity is needed to work out the structure of the neutron star. Calculations by Landau[54] and by Oppenheimer and Volkoff[55] showed that there is a maximum mass, beyond which neutron stars become unstable. Oppenheimer and Volkoff found this maximum mass to be $0.7 M_\odot$. But these calculations treated the neutrons as an ideal gas. For $\rho(0)$ greater than ρ_c, the separation (1.10.24) of neutrons is no greater than the range of nuclear forces, and a treatment of neutrons as free particles is no longer reliable. There have been various estimates of the maximum mass of stable neutron stars, all of the order of a few solar masses at most.

Because of their small size, neutron stars would naturally be expected to spin very rapidly. The Sun, with a radius $R_\odot$ of about 7×10^5 km, rotates with a frequency $\omega_\odot/2\pi = 5 \times 10^{-7}$ revolutions per second, not an exceptionally rapid rotation. If a progenitor star core had a similar radius and rotation rate, and if angular momentum per mass $\propto \omega R^2$ were conserved in its collapse to a neutron star, then the decrease in its radius to a few kilometers would increase its rate of rotation by a factor of order 10^{10}, giving it a rotation rate of a few thousand revolutions per second. This is roughly the maximum possible rotation rate. A body of mass M and radius R that is held together only by gravitation cannot rotate at an angular frequency ω greater than $\omega_{\max} \approx \sqrt{GM/R^3}$, at which rate the centripetal acceleration $\omega^2 R$ equals the gravitational acceleration GM/R^2. For a neutron star at the Oppenheimer–Volkoff limit, $M \simeq 0.7 M_\odot$ and $R \simeq 10$ km, so its maximum rotation rate $\omega_{\max}/2\pi$ is about 1,600 revolutions per second.

The theoretical anticipation[56] of rapidly rotating neutron stars was borne out by the discovery of pulsars. First came the observation[57] in 1967 of a source of radio pulses with period 1.33 seconds. It was proposed[58] that this was a rapidly rotating neutron star, emitting radiation along the direction of a strong magnetic field, at an angle to the axis of rotation, which happens to point in the direction of Earth once in each rotation. This suggestion became widely

---

[54] L. D. Landau, *Phys. Z. Sowjetunion* **1**, 285 (1933).
[55] J. R. Oppenheimer and G. M. Volkoff, *Phys. Rev.* **55**, 374 (1938).
[56] F. Pacini, *Nature* **216**, 567 (1967).
[57] A. Hewish, S. J. Bell, J. D. H. Pilkington, P. F. Scott, and R. A. Collins, *Nature* **217**, 709 (1968).
[58] T. Gold, *Nature* **218**, 731 (1968).

accepted with the discovery of a source of much more rapid pulses, with period 33 milliseconds, too rapid for anything but a neutron star, in a known supernova remnant, the Crab nebula. Since then pulsars have been found emitting radiation at various wavelengths, with pulse periods ranging from 8.5 seconds down to 1.4 milliseconds. There is uncertainty about the mechanism for producing this radiation, but there seems no doubt that the sources are rapidly rotating neutron stars.

Since the discovery of pulsars neutron stars have become even more interesting. As discussed in Section 2.3, pulsars were found in binary systems, whose decay gave the first observational evidence for the emission of gravitational radiation, and the coalescence of binary neutron stars was proposed to account for observed bursts of electromagnetic radiation (so-called *kilonovae*) with intrinsic luminosity between ordinary novae and supernovae. Most dramatically, as we shall see in Section 2.4, in 2017 gravitational waves as well as electromagnetic radiation were observed coming from the coalescence of a binary of neutron stars.

1.11 Supermassive Stars

There is an interesting class of stars in which the pressure of material particles is much less than radiation pressure, though not entirely negligible. As we shall see, these stars are necessarily supermassive, typically heavier than $100 M_\odot$. Stars this massive are very rare in the present universe,[59] but they are plausible precursors of supernovae that have led to neutron stars or black holes.

The energy density and pressure of radiation are given by the well-known formulas

$$\mathcal{E}_{\rm rad} = a T^4, \quad p_{\rm rad} = \frac{1}{3} a T^4, \qquad (1.11.1)$$

where a is the radiation energy constant

$$a = \frac{\pi^2 k_B^4}{5 \hbar^3 c^3} = 7.567 \times 10^{-15} \frac{\rm erg}{\rm cm^3\, K^4}. \qquad (1.11.2)$$

The matter of the star is assumed to form a non-relativistic ideal gas of particles of average mass $\overline{m}$, in thermal equilibrium with the radiation, and with energy density and pressure

$$\mathcal{E}_{\rm gas} = \frac{\rho k_B T}{\overline{m}(\gamma - 1)}, \quad p_{\rm gas} = \frac{\rho k_B T}{\overline{m}}, \qquad (1.11.3)$$

---

[59] One famous example is η Carinae A, the heavier star in a binary at a distance of 2,300 pc, with a mass estimated as $(100–200) M_\odot$. This is not a stable star; in 1837 it became the second brightest star in the sky, then faded to below naked-eye visibility, and in 1940 again became easily visible. It is estimated to have lost a mass of about $10 M_\odot$ in a decade.

where k_B is the Boltzmann constant, ρ is the mass density, γ is the polytrope index of the gas alone, not counting the radiation, and $\overline{m}$ is the mean mass of the gas particles. (For ionized hydrogen $\overline{m} \simeq m_p$, the proton mass, and $\gamma = 5/3$.) The ratio of gas pressure to radiation pressure is then

$$\beta \equiv p_{\text{gas}}/p_{\text{rad}} = \frac{3k_B}{a\overline{m}} \frac{\rho}{T^3}. \tag{1.11.4}$$

For $\beta \ll 1$ Eq. (1.11.1) shows that the star is close to a polytrope with index $\Gamma = 4/3$, so that $p \simeq K\rho^{4/3}$ with a constant K that can be expressed it terms of β:

$$K \simeq p_{\text{rad}}/\rho^{4/3} = \frac{aT^4}{3\rho^{4/3}} = \left(\frac{3}{a}\right)^{1/3} \left(\frac{k_B}{\overline{m}\beta}\right)^{4/3}. \tag{1.11.5}$$

Assuming the whole star to be in a state of effective convection, the constant K and hence also β must be constant through the star. This can be seen more quantitatively from considerations of entropy. As we saw in Section 1.7, the entropy per gram of an ideal gas with $\mathcal{E}_{\text{gas}} = p_{\text{gas}}/(\gamma - 1)$ is $R/(\gamma - 1) \ln\left(T\rho^{1-\gamma}\right)$, where $R = k_B/\overline{m}$ is the constant appearing in the ideal gas law $p_{\text{gas}} = R\rho T$. The entropy of the radiation per mass of gas is calculated from

$$T \, ds_{\text{rad}} = d\left(\frac{aT^4}{\rho}\right) + \frac{aT^4}{3} d\left(\frac{1}{\rho}\right),$$

from which it follows that $s_{\text{rad}} = 4aT^3/3\rho$. The total specific entropy is then

$$s = 4aT^3/3\rho + R/(\gamma - 1) \ln\left(T\rho^{1-\gamma}\right) = R\left[\frac{4}{\beta} + \frac{1}{\gamma - 1} \ln\left(p_{\text{gas}}/\rho^\gamma\right)\right].$$

We expect the logarithm to vary only by amounts of order unity (at least away from the star's surface), so in order for s to be constant $1/\beta$ can only change by amounts of order unity, and therefore for $\beta \ll 1$ by only a small fractional amount, at most of order β.

Using the general formula (1.8.8) for the mass of any non-relativistic polytrope, and setting $\Gamma = 4/3$, we have here

$$M = 4\pi \times 2.01824 \times \left(\frac{K}{\pi G}\right)^{3/2} = 18 M_\odot \left(\frac{m_p}{\overline{m}}\right)^2 \frac{1}{\beta^2}. \tag{1.11.6}$$

With $\overline{m} \simeq m_p/2$ and β less, say, than 0.3, the mass must be above $800 M_\odot$, and hence such stars are truly supermassive.

It should be noted that, even though we are interested here in the case $\beta \ll 1$, for which gas pressure is much less than radiation pressure, we can (and will) nevertheless confine our attention to the case $\rho c^2 \gg aT^4$, for which gas rest energy density is much greater than radiation energy density. The ratio is

$$\frac{\rho c^2}{aT^4} = \beta \frac{\overline{m} c^2}{3k_B T}.$$

1.11 Supermassive Stars

Hence, as long as the gas itself is non-relativistic, with $3k_B T \ll \overline{m}c^2$, we can assume that $\rho c^2 \gg aT^4$, provided only that β is greater than a lower bound $3k_B T/\overline{m}c^2$. For this reason, general relativity has so far played no role in our remarks about supermassive stars.

But, as shown in Section 1.9, in order to identify the critical density at which the star becomes unstable, we must find the stationary point of the internal energy E, which general relativity gives as the expression (1.9.2). To calculate the first term in Eq. (1.9.2), we need

$$\Delta \mathcal{E} \equiv \mathcal{E}_{\text{rad}} + \mathcal{E}_{\text{gas}} - 3p_{\text{rad}} - 3p_{\text{gas}} = -\frac{3\gamma - 4}{\gamma - 1} p_{\text{gas}} \simeq -\frac{3\gamma - 4}{\gamma - 1} \beta p. \quad (1.11.7)$$

The factor β makes this small, so we can calculate its integral over the star by taking the star to be a polytrope with $\Gamma = 4/3$, the corrections to this being doubly small. Making the appropriate substitutions (1.8.3) and (1.8.4) for a $\Gamma = 4/3$ polytrope, we have

$$\int 4\pi r^2 p \, dr = 4\pi K \rho^{4/3}(0) \left[\rho^{1/3}(0)(\pi G/K)^{1/2} \right]^{-3} \alpha, \quad (1.11.8)$$

where α is another numerical constant,

$$\alpha \equiv \int_0^{\xi_1} \Theta^4(\xi) \xi^2 \, d\xi = 1.18119. \quad (1.11.9)$$

Combining Eqs. (1.9.2), (1.11.7), and (1.11.8), we have then

$$E = -\frac{4\pi \beta K^{5/2} \alpha}{(\pi G)^{3/2}} \left(\frac{3\gamma - 4}{\gamma - 1} \right) \rho^{1/3}(0) + \frac{16\pi K^{7/2} \eta}{(\pi G)^{3/2} c^2} \rho^{2/3}(0). \quad (1.11.10)$$

The star is stable for sufficiently small central densities, where the second term, due to general relativistic corrections, can be neglected. With increasing central density, the star becomes unstable at a critical value ρ_{inst} of the central density at which $\partial E/\partial \rho(0) = 0$:

$$\rho_{\text{inst}} = \left[\frac{\beta \alpha c^2}{8 K \eta} \frac{3\gamma - 4}{\gamma - 1} \right]^3. \quad (1.11.11)$$

To bring out the physical significance of this result, it is useful to consider the radius of the star, which according to Eq. (1.8.7) is given by $R = \xi_1 (K/\pi G)^{1/2} \rho^{-1/3}(0)$, where $\xi_1 = 6.89685$ is the value of ξ where $\Theta(\xi)$ drops to zero. At the critical central density, this is

$$R_{\text{inst}} = \frac{8\eta \xi_1}{\beta \alpha c^2} (\pi G)^{-1/2} K^{3/2} \frac{\gamma - 1}{3\gamma - 4}. \quad (1.11.12)$$

It is instructive to compare this with the Schwarzschild radius $2MG/c^2$. The mass of a $\Gamma = 4/3$ polytrope is given by Eq. (1.11.6), so

$$\frac{c^2 R_{\text{inst}}}{MG} = \frac{\gamma - 1}{\beta(3\gamma - 4)} \frac{2 \times 6.89685 \times 3.49815}{2.01824 \times 1.18119} = 20.24 \frac{\gamma - 1}{\beta(3\gamma - 4)}. \quad (1.11.13)$$

For a stable supermassive star we have $R > R_{\text{inst}}$, so for small β the star's radius is much larger than the Schwarzschild radius, and the redshift MG/Rc^2 from the surface of the star is quite small.

Finally, let us consider the evolution of a supermassive star. The mass M of the star is dominated by the rest mass of the nucleons it contains, and hence cannot appreciably change with time. According to Eq. (1.11.6), it follows that K does not change much, so the same is true of β. But the central density can and does evolve. The internal energy (1.11.10) may be written

$$E(\rho(0)) = E_0 \left[-2 \left(\frac{\rho(0)}{\rho_{\text{inst}}} \right)^{1/3} + \left(\frac{\rho(0)}{\rho_{\text{inst}}} \right)^{2/3} \right], \quad (1.11.14)$$

where E_0 is a positive constant. If initially a supermassive star has $\rho(0) < \rho_{\text{inst}}$ it will have an internal energy that decreases monotonically with central density, and be stable. As the star radiates, it loses internal energy, which must then become increasingly negative. Since for $\rho(0) < \rho_{\text{inst}}$ the internal energy E is a monotonically decreasing function of central density, this requires the central density to rise, until $\rho(0)$ reaches its critical value, whereupon the star explodes. It is not clear what happens after that, but it is plausible that a stable remnant is left, a star that is no longer supermassive.

Bibliography for Chapter 1

- E. Böhm-Vitesse, *Stellar Astrophysics* (Cambridge University Press, Cambridge, 1989)
  - Volume 1: *Basic Stellar Observations and Data*,
  - Volume 2: *Stellar Atmospheres*,
  - Volume 3: *Stellar Structure and Evolution*.
- S. Chandrasekhar, *An Introduction to the Theory of Stellar Structure* (University of Chicago Press, Chicago, IL, 1939).
- H.-Y. Chiu, *Stellar Physics* (Blaisdell, Waltham, MA, 1968).
- J. P. Cox and R. T. Giuli, *Principles of Stellar Structure* (Gordon & Breach, New York, 1968).
- C. J. Hansen, S. D. Kawaler, and V. Trimble, *Stellar Interiors*, 2nd edn. (Springer, New York, 2004).

- B. K. Harrison, K. S. Thorne, M. Wakano, and J. A. Wheeler, *Gravitation Theory and Gravitational Collapse* (University of Chicago Press, Chicago, IL, 1965).
- R. Kippenhahn and A. Weigert, *Stellar Structure and Evolution* (Springer-Verlag, Berlin, 1990).
- F. LeBlanc, *Introduction to Stellar Atmospheres* (John Wiley & Sons, Chichester, 2010).
- D. Prialnik, *Theory of Stellar Structure and Evolution*, 2nd edn. (Cambridge University Press, Cambridge, 2010).
- G. B. Rybicki and A. Lightman, *Radiative Processes in Astrophysics* (John Wiley & Sons, New York, 1979).
- M. Schwarzschild, *Structure and Evolution of the Stars* (Princeton University Press, Princeton, NJ, 1958).
- S. Shapiro and S. Teukolsky, *Black Holes, White Dwarfs, and Neutron Stars* (Wiley, New York, 1983).
- F. H. Shu, *The Physics of Astrophysics – Volume 1 Radiation* (University Science Books, Mill Valley, CA, 1991).
- R. J. Tayler, *The Stars: Their Structure and Evolution* (Wykeham Publications, London, 1970).
- A. Weiss, W. Hildebrandt, H.-C. Thomas, and H. Ritter, *Cox and Giuli's Principles of Stellar Structure* (Cambridge Scientific Publishers, Cambridge, 2004).

2
Binaries

We have been considering stars in isolation, but a good fraction of the stars in our galaxy are binaries, pairs of stars in orbit about a common center of mass. Binary stars have long provided invaluable information about individual stars, such as the measurements of stellar masses mentioned in Section 1.6. We begin here in Section 2.1 with a reminder of the Newtonian theory of these orbits, and discuss how the stellar masses and orbital parameters can be measured. In Section 2.2 we take up a special class of binaries, stars so close in their orbits that matter can spill from one onto the other. Binaries gained new importance with the discovery of a binary pulsar, which as described in Section 2.3 showed a loss of energy that provided the first observational evidence for the existence of gravitational radiation. Section 2.4 tells how the excitement was heightened in 2015 with the first actual detection of gravitational waves, produced by a binary of black holes caught at the moment of coalescence. Since then gravitational waves have been detected from the coalescence of several other binaries of black holes and of neutron stars.

2.1 Orbits

Let us first work out the motion of the stars in a general binary, and how it is observed. Taking the origin of the coordinate system as the center of mass, the coordinate vectors $\mathbf{x}_1$ and $\mathbf{x}_2$ of the two stars are related by $m_1 \mathbf{x}_1 + m_2 \mathbf{x}_2 = 0$, where m_1 and m_2 are the two stars' masses. Both coordinate vectors can thus be expressed in terms of the separation $\mathbf{r} \equiv \mathbf{x}_1 - \mathbf{x}_2$, as

$$\mathbf{x}_1 = \left(\frac{m_2}{M}\right) \mathbf{r}, \qquad \mathbf{x}_2 = -\left(\frac{m_1}{M}\right) \mathbf{r}, \qquad (2.1.1)$$

with $M \equiv m_1 + m_2$ the total mass. For stars moving much more slowly than light, we can use the Newtonian equations of motion:

$$\frac{d^2}{dt^2}\mathbf{x}_1 = -Gm_2 \mathbf{r}/r^3, \qquad \frac{d^2}{dt^2}\mathbf{x}_2 = +Gm_1 \mathbf{r}/r^3.$$

Both equations can more conveniently be written as

$$\frac{d^2}{dt^2}\mathbf{r} = -GM\mathbf{r}/r^3. \qquad (2.1.2)$$

We take the 3-axis to be normal to the orbital plane, and write the solution as

$$\mathbf{r} = r\left(\cos\varphi, \sin\varphi, 0\right), \qquad (2.1.3)$$

where

$$r = \frac{L}{1 - e\cos\varphi} \qquad (2.1.4)$$

and

$$\frac{d\varphi}{dt} = \sqrt{\frac{GM}{L^3}}(1 - e\cos\varphi)^2. \qquad (2.1.5)$$

Here e and L are elements of the elliptical orbit known as the eccentricity and semi-latus rectum. For an ellipse with major axis $2a$ and minor axis $2b$, these elements are $e = \sqrt{1 - b^2/a^2}$ and $L = a(1 - e^2)$. Our coordinates are here chosen so that the major axis of the ellipse is along the 1-axis, with apastron at $\varphi = 0$, where r takes its maximum value $a(1+e)$, and with periastron at $\varphi = \pi$, where r reaches its minimum $a(1-e)$.

The motion is periodic, with period

$$T = \int_0^{2\pi} \frac{d\varphi}{d\varphi/dt} = \sqrt{\frac{L^3}{MG}} \int_0^{2\pi} \frac{d\varphi}{(1 - e\cos\varphi)^2} = 2\pi\sqrt{\frac{a^3}{MG}}. \qquad (2.1.6)$$

For future reference, we note that the total kinetic and potential energy of the stars is

$$E = m_1 \frac{\dot{\mathbf{x}}_1^2}{2} + m_2 \frac{\dot{\mathbf{x}}_2^2}{2} - \frac{Gm_1m_2}{r}$$

and their angular momentum is

$$\mathbf{J} = m_1 \mathbf{x}_1 \times \dot{\mathbf{x}}_1 + m_2 \mathbf{x}_2 \times \dot{\mathbf{x}}_2.$$

The angular momentum is a vector normal to the orbit, with magnitude $J_\perp$. Using Eqs. (2.1.1) and (2.1.3)–(2.1.5), we find that the energy and angular momentum are

$$E = -\frac{\mu MG}{2a}, \quad J_\perp = \mu\sqrt{GML}, \qquad (2.1.7)$$

where $\mu \equiv m_1 m_2/(m_1 + m_2)$ is the usual reduced mass.

Now let us consider how observations are used to find the properties of binary stars. Some, like Sirius, are visual binaries, pairs of stars that are far enough apart and close enough to Earth that the stars can be separately observed visually. Because of their large separation, they often move too slowly to allow an accurate measurement of their velocities. Much more revealing are eclipsing variables, whose line of sight to the Earth happens to lie in the binary star orbital plane, so that each star periodically eclipses the other. As mentioned in Section 1.6, the analysis of their light curves, of luminosity vs. time, has provided values for the masses and radii of a large sample of stars, but this is a complicated business that will not be pursued here. We will be chiefly concerned in what follows with the many spectroscopic binaries, whose properties are generally known only through Doppler shifts of spectral lines or pulses, due to orbital motion. This class includes the Hulse–Taylor binary pulsar, discussed in Section 2.3, which provided the first observational evidence for the reality of gravitational radiation.

First, let us calculate the Doppler shift of any periodic signal from either of the two stars. We take the line of sight from the binary system to the Earth in the coordinate system used in Eq. (2.1.3) to be the general unit vector

$$\mathbf{n} = (\sin i \, \cos \psi, \sin i \, \sin \psi, \cos i), \qquad (2.1.8)$$

with i the angle between the line of sight and the normal to the orbit. The motion of the two stars produces Doppler shifts

$$(\Delta \nu/\nu)_1 = \mathbf{n} \cdot \dot{\mathbf{x}}_1/c = (m_2/M)\mathbf{n} \cdot \dot{\mathbf{r}}/c,$$
$$(\Delta \nu/\nu)_2 = \mathbf{n} \cdot \dot{\mathbf{x}}_2/c = -(m_1/M)\mathbf{n} \cdot \dot{\mathbf{r}}/c. \qquad (2.1.9)$$

Using the Newtonian results (2.1.4) and (2.1.5), we have

$$\dot{\mathbf{r}} = \sqrt{\frac{MG}{Lc^2}}(-\sin\varphi, \, \cos\varphi - e, \, 0),$$

so, in the absence of relativistic effects, the Doppler shifts are

$$(\Delta\nu/\nu)_1 = (m_2/M)\sqrt{\frac{MG}{Lc^2}} \sin i \left[\sin(\psi - \varphi) - e \sin\psi \right],$$
$$(\Delta\nu/\nu)_2 = -(m_1/m_2)(\Delta\nu/\nu)_1. \qquad (2.1.10)$$

The comparison of Eq. (2.1.10) with the observed variation with time of the Doppler shift from either star together with Eq. (2.1.5) allows the determination of values both for MG/L^3 and for the eccentricity e. Where both Doppler shifts are observed, their ratio also gives the mass ratio m_1/m_2. The overall scale of the Doppler shift for star 1 can yield a value for $(m_2/M)\sqrt{MG/L} \sin i$, and if both Doppler shifts are measured m_2/M will be known, so this gives a value for $\sqrt{MG/L} \sin i$. But without further data this does not yield separate values for M, L, or $\sin i$.

Fortunately, the additional data needed to find the properties of the binary can come from relativistic corrections. One general relativistic phenomenon that affects Doppler shifts is the precession of the elliptical orbit of $\mathbf{r}$. The precession in radians per revolution is given by the same formula used by Einstein in 1914–1915 to calculate the precession of the orbit of Mercury,[1]

$$\Delta\varphi = \frac{6\pi MG}{Lc^2},$$

but with M now of course interpreted as the total mass of the binary system, rather than of the Sun. In Eq. (2.1.8), ψ is the angle between the major axis of the ellipse and the projection of the line of sight from the binary to the solar system on the orbital plane, so ψ increases by the angle $\Delta\varphi$ per revolution, producing a non-periodic secular change in the Doppler shifts (2.1.10). This change depends on MG/L, so with the value of MG/L^3 found from the observed period, it is possible to make a separate determination of MG and L, but this still does not give separate values for m_1 and m_2 if Doppler shifts are observed from only one star of the binary.

The observed Doppler shift of either star is also affected by a relativistic time dilation, due both to the gravitational field of the companion star and to the motion of the observed star. As shown in the appendix to this section, this produces an added frequency shift. For star 1,

$$(\Delta\nu/\nu)_{\text{Einstein},1} = \text{constant} - (m_2 G/MLc^2)(M+m_2)e\cos\varphi. \qquad (2.1.11)$$

With M, e, and L already known, the presence of this term in the observed frequency shift allows a separate determination of m_1 and m_2. There is also a slowing of signals from one star in the gravitational field of the companion star, analogous to the Shapiro time delay[2] in the gravitational field of the Sun.

Appendix: Calculation of Time Dilation in Binary Stars

If $\delta\tau = 1/\nu$ is the proper time interval between emitted wave crests or pulses, measured in the co-moving inertial frame of the emitting star, and δt is the coordinate time between these wave crests or pulses, in a coordinate system in which the center of mass is at rest, then

$$\delta\tau = \delta t \sqrt{-g_{00} - 2g_{0i}v_{1i} - g_{ij}v_{1i}v_{1j}}, \qquad (2.1.\text{A1})$$

where $\mathbf{v}_1$ is the velocity of the observed star relative to the center of mass, and $g_{\mu\nu}$ are the components of the metric tensor in a coordinate system that is Minkowskian far from the binary. According to the post-Newtonian approximation to general relativity,[3] as long as the emitting star's velocity is much

---
[1] A. Einstein, *Sitz. preuß. Acad. Wiss.* 1914, p. 1030; 1915, pp. 778, 799, 831, 844. For a textbook derivation, see Section 8.6 of S. Weinberg, *Gravitation and Cosmology* (Wiley, New York, 1972).
[2] I. I. Shapiro, *Phys. Rev. Lett.* **13**, 789.
[3] For a textbook treatment, see S. Weinberg, *Gravitation and Cosmology* (Wiley, New York, 1972), Section 9.1. In this appendix, we use units with $c = 1$.

less than that of light, we have $g_{00} = -1 - 2\phi + O(\bar{v}^4)$, $g_{0i} = O(\bar{v}^3)$, and $g_{ij} = \delta_{ij} + O(\bar{v}^2)$, where ϕ is the gravitational field of the companion at the position of the emitting star. Hence to order $\bar{v}^2$ we have

$$\delta\tau = \delta t \sqrt{1 + 2\phi - \mathbf{v}_1^2} = \delta t (1 + \phi - \mathbf{v}_1^2/2), \qquad (2.1.A2)$$

the corrections to this formula being of order $\bar{v}^4$. Now,

$$\phi = -Gm_2/r = -(Gm_2/L)(1 - e\cos\varphi)$$

and

$$\mathbf{v}_1^2 = (m_2/M)^2 \, \dot{\mathbf{r}}^2 = (m_2/M)^2 (MG/L)\left(1 - 2e\cos\varphi + e^2\right),$$

so there is an "Einstein" fractional frequency shift

$$\begin{aligned}(\Delta\nu/\nu)_{\text{Einstein},1} &= (\delta\tau/\delta t)^{-1} - 1 = -\phi + \mathbf{v}_1^2/2 \\ &= (m_2 G/L)(1 - e\cos\varphi) \\ &\quad + (m_2/M)^2 (MG/2L)\left(1 - 2e\cos\varphi + e^2\right). \end{aligned} \qquad (2.1.A3)$$

A constant fractional frequency shift is usually undetectable, since we generally have no knowledge of what ν would be in the absence of these relativistic effects and the Doppler shift, so the interesting part of this frequency shift is its variable part, reported in Eq. (2.1.11).

2.2 Close Binaries

In the orbital analysis of the previous section the two stars in a binary were treated as point masses. Now we will consider binaries in which the radius of one or both of the stars is not much smaller than the separation of the binary, and we have to take into account the distortion of the star or stars by tidal effects. We can still give an analytic treatment, using the same Newtonian results for orbital motion, in the common case where most of each star's mass remains in a sphere but the outer layer of at least one star is severely distorted, to the point where matter can flow from that star to the other.

We begin by asking how far the stars in a binary need to be separated for tidal distortion to be unimportant, and will then turn to the case when they are not that far apart. Consider one of the stars in a binary, with mass m_1 and radius R_1, in a circular orbit about the center of mass with circular frequency Ω, at a distance r from the other star. A particle of mass δm on the surface of star 1 at the point closest to the other star will feel a gravitational force toward the center of star 1:

$$F_{\text{grav}} = \frac{Gm_1 \, \delta m}{R_1^2} - \frac{Gm_2 \, \delta m}{(r - R_1)^2}, \qquad (2.2.1)$$

the minus sign preceding the second term indicating that this part of the force is toward the other star. Since the particle is at a distance $m_2 r/M - R_1$ from the center of mass of the two stars, it will also feel a centrifugal force in the same direction:[4]

$$F_{\text{cent}} = \Omega^2 (m_2 r/M - R_1) \, \delta m. \qquad (2.2.2)$$

(We are here assuming that the centrifugal force from any rotation of the star is much less than from its orbital motion.) We recall from Eq. (2.1.6) that for circular orbits with $a = r$, the angular frequency $\Omega = 2\pi/T$ has $\Omega^2 = MG/r^3$, so that the gravitational and centrifugal forces would balance each other for $m_1 = 0$ and $R_1 \to 0$. For $R_1 \neq 0$ the centrifugal force (2.2.2) is less than the gravitational force exerted by the other star, and so the particle would fly off the surface of star 1 if not held by the gravity of star 1. The condition for particles on the surface of star 1 not to fly off toward star 2 is thus that

$$\frac{Gm_1}{R_1^2} - \frac{Gm_2}{(r-R_1)^2} + \Omega^2 (m_2 r/M - R_1) > 0,$$

or, using $\Omega^2 = MG/r^3$ and canceling factors of G,

$$\frac{m_1}{R_1^2} - \frac{m_2}{(r-R_1)^2} > \frac{MR_1 - m_2 r}{r^3}.$$

This is a necessary condition for neglecting tidal distortion of the side of star 1 facing toward star 2, and it is reasonable to suppose that the equality of the two sides of this condition defines a value of r at which tidal distortion begins to matter, but strictly speaking, since this condition was derived assuming both stars to be spheres, all we can say with confidence is that these tidal effects are negligible if

$$\frac{m_1}{R_1^2} - \frac{m_2}{(r-R_1)^2} \gg \frac{MR_1 - m_2 r}{r^3}. \qquad (2.2.3)$$

Similarly, a sufficient condition for neglecting tidal distortion of the side of star 1 facing away from star 2 is that

$$\frac{m_1}{R_1^2} + \frac{m_2}{(r+R_1)^2} \gg \frac{MR_1 + m_2 r}{r^3}. \qquad (2.2.4)$$

In the common case where r is much larger than the stellar radius R_1, both Eqs. (2.2.3) and (2.2.4) amount to the requirement that

$$r \gg R_1 \left(1 + \frac{3m_2}{m_1}\right)^{1/3}. \qquad (2.2.5)$$

---

[4] For reasons I am not able to understand, the textbook treatments of this problem that I have seen neglect the decrease in centrifugal force given by the term $-R_1$ in parentheses in Eq. (2.2.2). As a result, Eq. (2.2.5) is generally given with a factor 2 in place of 3 in the cube root.

This is known as the Roche limit, named for Édouard Albert Roche (1820–1883).

For instance, Sirius is a famous visual binary consisting of the stars Sirius A and B with masses $M_A = 2.28 M_\odot$ and $M_B = 0.98 M_\odot$ and radii $R_A = 1.71 R_\odot$ and $R_B = 0.008 R_\odot$, respectively. (Sirius B is a white dwarf.) According to condition (2.2.5), the tidal distortion of Sirius A will be negligible if it is at a distance from Sirius B much greater than $1.3 R_A = 2.3 R_\odot$. Since its actual distance to Sirius B is 1.2×10^9 km $= 1.7 \times 10^3 R_\odot$, this condition is very well satisfied. Sirius B is so small that tidal effects on it are even less than for Sirius A.

Where Eqs. (2.2.3) and (2.2.4) are not satisfied, we may need to take into account the tidal distortion of the stars' shapes. If the binary is in rigid rotation with angular frequency Ω around an axis with direction $\hat{n}$ through the center of mass, then the gravitational plus centrifugal force on a test body of mass δm at position $\mathbf{x}$ is

$$\mathbf{F}(\mathbf{x}) = \delta m \left(-\nabla \phi(\mathbf{x}) + \Omega^2 [\mathbf{x} - \hat{n}(\hat{n} \cdot \mathbf{x})] \right),$$

where $\phi(\mathbf{x})$ is the total gravitational potential. This force can be written as the gradient of an effective potential,

$$\mathbf{F}(\mathbf{x}) = -\delta m \, \nabla \Phi(\mathbf{x}),$$

where

$$\Phi(\mathbf{x}) = \phi(\mathbf{x}) - \frac{\Omega^2}{2} \left[\mathbf{x}^2 - (\hat{n} \cdot \mathbf{x})^2 \right].$$

This force can be balanced by pressure forces in the interior of the stellar material, and on the surface its normal component can also be balanced by pressure forces, but there is nothing to balance tangential components of the force on the surface, so the tangential components of $\nabla \Phi$ must vanish on the surface, and therefore all points on the surface of the stellar material must have the same values for $\Phi(\mathbf{x})$:

$$\phi(\mathbf{x}) - \frac{\Omega^2}{2} \left(\mathbf{x}^2 - (\hat{n} \cdot \mathbf{x})^2 \right) = \Phi, \tag{2.2.6}$$

where Φ is now a constant, whose various possible values define various possible equipotential surfaces that could be the boundary of stellar matter.

Equation (2.2.6) is difficult to use in cases where tidal effects disrupt the whole mass distribution, because $\phi(\mathbf{x})$ then depends on the shape of this surface. But we are chiefly concerned with tidal effects on the outer layers of the stars in a binary, where most of the masses of these stars remain in nearly spherically symmetric configurations centered on positions $\mathbf{x}_1$ and $\mathbf{x}_2$, in which case we can approximate the gravitational potential as

2.2 Close Binaries

$$\phi(\mathbf{x}) = -\frac{m_1 G}{|\mathbf{x}-\mathbf{x}_1|} - \frac{m_2 G}{|\mathbf{x}-\mathbf{x}_2|}, \qquad (2.2.7)$$

and again take $\Omega^2 = MG/r^3$, so that Eq. (2.2.6) becomes

$$\frac{m_1 G}{|\mathbf{x}-\mathbf{x}_1|} + \frac{m_2 G}{|\mathbf{x}-\mathbf{x}_2|} + \frac{MG}{2r^3}\left(\mathbf{x}^2 - (\hat{n}\cdot\mathbf{x})^2\right) = -\Phi. \qquad (2.2.8)$$

Taking the centers of the spherical mass distributions in rotating Cartesian coordinates to be at fixed positions

$$\mathbf{x}_1 = (-m_2 r/M, 0, 0), \quad \mathbf{x}_2 = (m_1 r/M, 0, 0),$$

and the rotation axis to be in the direction $\hat{n} = (0,0,1)$, condition (2.2.8) requires that at a position $\mathbf{x} = (x,y,z)$ we have

$$\frac{m_1 G}{((x+m_2 r/M)^2 + y^2 + z^2)^{1/2}} + \frac{m_2 G}{((x-m_1 r/M)^2 + y^2 + z^2)^{1/2}}$$
$$+ \frac{MG}{2r^3}(x^2+y^2) = -\Phi. \quad (2.2.9)$$

There is an equipotential surface for each constant negative Φ. Of course, these mathematical surfaces may or may not actually be filled with stellar material.

For $-\Phi$ sufficiently large, the equipotential surface consists of disconnected parts,[5] two small spherical surfaces around the positions $\mathbf{x}_1$ and $\mathbf{x}_2$, with radii proportional respectively to m_1 and m_2, and to $1/|\Phi|$. For smaller values of $-\Phi$ these spheres are larger, and distorted by tidal effects, with the surfaces being pulled toward one another. For some critical value of Φ the two surfaces meet at a point (known as L1) on the line between $\mathbf{x}_1$ and $\mathbf{x}_2$. The volumes enclosed by the equipotential surfaces around $\mathbf{x}_1$ and $\mathbf{x}_2$ for this critical value of Φ are known as Roche lobes.

If one star fills its Roche lobe (as may be the case for main sequence stars and red giants) and the other does not (as is generally true of compact stars such as white dwarfs and neutron stars) then pressure forces will push matter from the filled lobe to the unfilled one. One effect of this transfer of matter is to produce radiation as matter falls on the compact star, usually at X-ray frequencies. Another effect is to speed up or slow down the rotation of the binary. Using Eq. (2.1.3) with $e = 0$ gives $\Omega^2 = GM/r^3$, so (ignoring any

---

[5] As long as the equipotential surface consists of disconnected parts, we could take the left-hand side of Eq. (2.2.9) to have different values on each disconnected part of the surface. Indeed, for real binaries with a separation large enough to satisfy condition (2.2.4), these values are generally different for the actual surfaces of each star. We are taking (2.2.9) to have equal values for each disconnected part of the equipotential surface because we are now chiefly interested in the case where the inequality (2.2.4) is *not* satisfied, and where these parts of the equipotential surface merge and become a single connected surface, for which of course there is just a single value of Φ.

possible rotation of the individual stars), we see that the angular momentum of the binary is

$$J = \Omega m_1 \left(\frac{m_2}{M}r\right)^2 + \Omega m_2 \left(\frac{m_1}{M}r\right)^2 = \frac{m_1 m_2 \Omega r^2}{M} = \frac{m_1 m_2 G^{2/3}}{\Omega^{1/3} M^{1/3}}. \quad (2.2.10)$$

The conservation of mass requires that $\dot{m}_1 = -\dot{m}_2$, so with Eq. (2.2.10) the conservation of angular momentum gives

$$\frac{\dot{\Omega}}{\Omega} = \frac{3\dot{m}_1(m_2 - m_1)}{m_1 m_2}. \quad (2.2.11)$$

Hence, if we see Ω decreasing or increasing, we can infer that mass is being transferred from the more massive to the less massive star, or vice versa.

The transfer of mass from one star to another in a binary can have more dramatic effects. If an ordinary star that fills its Roche lobe is in a binary with a white dwarf, it will transfer mass to the dwarf. As the white dwarf mass increases, its central density increases, and may reach a value at which thermonuclear reactions ignite. This is a likely mechanism for producing the observed stellar explosions known as Type 1a supernovae, which are used as standard candles in measurements of the distances of galaxies with large red shifts. ("Type 1" simply indicates an absence of hydrogen in the spectrum of the supernova, which is consistent with the expected chemical composition of white dwarfs. The "a" distinguishes these supernovae from those that occur through the core collapse of massive stars that have exhausted their hydrogen, which have a very different curve of luminosity vs. time than supernovae of Type 1a.) There is some doubt about this explanation of Type 1a supernovae, because white dwarfs in binaries may lose enough mass through novae that they never experience thermonuclear explosions, and because no one has ever observed the companion stars from which white dwarfs in Type 1a supernovae could have acquired their mass. The question is open.

All this provides motivation to look in more detail at Roche lobes. To put Eq. (2.2.9) in dimensionless form, we can set $\Phi = -(MG/r)C$ and $(x, y, z) = r(x', y', z')$, so that

$$\frac{m_1/M}{((x' + m_2/M)^2 + y'^2 + z'^2)^{1/2}} + \frac{m_2/M}{((x' - m_1/M)^2 + y'^2 + z'^2)^{1/2}} + \frac{x'^2 + y'^2}{2} = C. \quad (2.2.12)$$

In the coordinates (x', y', z') the equipotential surface evidently depends only on C and on the mass ratio m_1/m_2. In particular, the critical value of C at which the surface first becomes connected as C decreases depends only on the mass ratio, as do the properties of the Roche lobes in these coordinates.

For general mass ratios the properties of the Roche lobes can be found quantitatively only by numerical calculation. However, it is possible to give an analytic treatment of the important case where one mass is much less than the other, say $m_2/m_1 \equiv \epsilon \ll 1$. As long as C is well above 3/2 the small spherical surface with center at $x' = 1 + O(\epsilon)$, $y' = z' = 0$ and radius $\simeq \epsilon/(C - 3/2)$ is part of the equipotential surface (2.2.12), since on this small sphere the first and third terms on the left-hand side of Eq. (2.2.12) have values respectively equal to 1 and 1/2 for $\epsilon = 0$. The other disconnected part of the equipotential surface will consist of points on which, apart from terms of order ϵ,

$$\frac{1}{(x'^2 + y'^2 + z'^2)^{1/2}} + \frac{x'^2 + y'^2}{2} = C,$$

since for C well above 3/2 all points on this surface are nowhere near the small sphere on which the second term in Eq. (2.2.12) is non-negligible. But as C decreases to the critical value $C = 3/2$, this surface approaches the small sphere, and the equipotential surface becomes connected. At this critical value, the equipotential surface becomes

$$\frac{1}{(x'^2 + y'^2 + z'^2)^{1/2}} + \frac{x'^2 + y'^2}{2} = 3/2, \tag{2.2.13}$$

except for points very close to $x' = 1$, $y' = z' = 0$, where the lobes are joined. This is an oblate figure of rotation about the z'-axis, with equatorial radius unity and poles at $x' = y' = 0$, $z' = \pm 2/3$. The volume of the Roche lobe around mass 1 is thus

$$V_1 = \pi r^3 \mathcal{I},$$

where $\mathcal{I}$ is the dimensionless integral

$$\mathcal{I} = \int_{-2/3}^{2/3} (x'^2 + y'^2) \, dz'.$$

We can use Eq. (2.2.13) to write

$$z'^2 = -\rho^2 + \frac{4}{(3 - \rho^2)^2},$$

where $\rho^2 \equiv x'^2 + y'^2$. The integral $\mathcal{I}$ can thus be written

$$\mathcal{I} = \int_0^1 \frac{\rho^2 (1 - 8(3 - \rho^2)^{-3})}{\sqrt{4(3 - \rho^2)^{-2} - \rho^2}} \, d\rho^2$$

$$= -\frac{16}{3} + 4\sqrt{3} + 8 \tanh^{-1}(2) - 8 \tanh^{-1}(\sqrt{3}) = 0.721487.$$

It is conventional to define an effective radius r_1, such that the volume of the Roche lobe around mass 1 is $4\pi r_1^3/3$. We see that in the case $m_1 \gg m_2$ the effective radius is $(3\mathcal{I}/4)^{1/3} r = 0.814886 r$. This is in perfect agreement with a numerical calculation of Kopal,[6] who gave $r_1 = 0.8149r$ for $m_1/m_2 \to \infty$. Also, we saw that the Roche lobe around mass 2 becomes infinitesimal for $m_1 \gg m_2$; likewise, the Roche lobe around mass 1 becomes infinitesimal for $m_1 \ll m_2$, so in this case $r_1 = 0$. The effective radii calculated numerically by Eggleton[7] for these and other mass ratios are given in a table below. Not too much emphasis should be put on the precision of these numerical results, as they are derived under the assumption that most of the masses of the stars remains in spherical distributions, which can at best be only a fair approximation.

m_1/m_2	r_1/r
∞	0.8149
1000	0.7817
100	0.7182
10	0.5803
2.5	0.4621
1	0.3799
0.4	0.3026
0.1	0.2054
0.01	0.1012
0.001	0.0482
0	0

2.3 Gravitational Wave Emission: Binary Pulsars

The orbital analysis outlined in Section 2.1 became dramatically more important with the discovery of a pulsar in a binary system. As discussed in Section 1.10, pulsars are rapidly rotating neutron stars, with masses typically comparable to the Sun's mass, with radii of order of a few kilometers, and with strong magnetic fields that are not parallel to the axis of rotation. The rotating neutron star produces a narrow beam of radio frequency electromagnetic waves along the direction of the magnetic field, which swivels around the axis of rotation. We on Earth receive a pulse of radiation when this beam happens to point toward us.

In 1974, in the course of a survey of pulsars using the Arecibo radio telescope in Puerto Rico, Russell Hulse and Joseph Taylor discovered a pulsar, PSR 1913+16, whose period of about 59 milliseconds varied periodically by

---

[6] Z. Kopal, *Close Binary Systems* (Chapman and Hall, London, 1959).
[7] P. A. Eggleton, *Astrophys. J.* **268**, 368 (1983). Eggleton's results provide small corrections to earlier calculations of Kopal, *op. cit.*, and B. Paczyński, *Ann. Rev. Astron. Astrophys.* **9**, 183 (1971).

about 0.07% over a period of about 8 hours.[8] This variation was interpreted as a Doppler shift, indicating that the pulsar and another invisible star were in an orbit about each other, with a period of about 8 hours, and a velocity component along the line of sight not much greater than 0.07% of the speed of light. Since the orbital velocity is thus much less than the speed of light, it is a good approximation to treat the orbit using Newtonian mechanics, as done in Section 2.1.

Fortunately, the observation of pulsar timing is so precise that it is possible to detect the relativistic effects discussed in Section 2.1, effects that depend on the parameters of the binary pulsar so that all of them can be calculated. When all these effects are put together, along with a decrease in orbital period taken as a free parameter, Weisberg, Nice, and Taylor in 2010 found that the least-squares fit of observed pulsar timing with theory yields the following parameters:[9]

- $m_1 = 1.4398(2) M_\odot$
- $m_2 = 1.3886(2) M_\odot$
- $e = 0.6171334(5)$
- $T = 0.32299744891(4)$ days
- $\dot{T} = -2.296(6) \times 10^{-12}$,

with the uncertainty in the last digit given for each item by the number in the parentheses. (A small part of the directly observed decrease in the orbital period is a kinematic effect, due to the acceleration of the binary system in the gravitational field of the galaxy. The value given above corrects for this effect.)

From the above values of m_1, m_2, and T, we can also calculate a value for the semi-major axis of the orbit of **r**:

$$a = (T/2\pi)^{2/3}(MG)^{1/3} = 1.95 \times 10^6 \text{ km}. \tag{2.3.1}$$

Thus the binary system could fit easily into the orbit of Mercury, which has semi-major axis 58 million kilometers. With these parameters, the rate of precession of the periastron is

$$\frac{\Delta\varphi}{T} = \frac{6\pi MG}{(1-e^2)c^5 aT} = 4.22°/\text{year}, \tag{2.3.2}$$

which is a huge rate when compared with the famous 43 *arc seconds per century* anomalous precession of the orbit of Mercury.

From the beginning it seemed likely that the observed speed up of the orbital motion of the binary pulsar is due to the emission of gravitational radiation. Just as accelerated electric charges produce electromagnetic radiation, so also accelerated masses produce gravitational radiation. Specifically, as is well known,

---

[8] R. A. Hulse and J. H. Taylor, *Astrophys. J.* **195**, L51 (1975).
[9] J. M. Weisberg, D. J. Nice, and J. H. Taylor, *Astrophys. J.* **722**, 1030 (2010), and earlier work of Taylor and Weisberg cited therein.

the acceleration of a set of particles with position vectors $\mathbf{x}_N$ and electric charge e_N radiates electromagnetic waves, with the average radiated power given in the dipole approximation by the Larmor formula[10]

$$\langle P \rangle = \frac{2}{3c^3} \left\langle \ddot{D}_i(t) \ddot{D}_i(t) \right\rangle,$$

where D_i is the electric dipole moment:

$$D_i(t) = \sum_N e_N x_{Ni}(t).$$

Here N runs over particle labels, and i runs over Cartesian coordinate indices 1, 2, 3, with repeated indices summed. Similarly, accelerated particles with masses m_N produce gravitational waves, with the average radiated power given in what is known as the quadrupole approximation by

$$\langle P \rangle = \frac{G}{5c^5} \left\langle \dddot{Q}_{ij}(t) \dddot{Q}_{ij}(t) - \frac{1}{3} \left| \dddot{Q}_{ii}(t) \right|^2 \right\rangle, \tag{2.3.3}$$

where Q_{ij} is the mass tensor

$$Q_{ij}(t) \equiv \sum_N m_N x_{Ni}(t) x_{Nj}(t). \tag{2.3.4}$$

In both cases, the average is over a time that is longer than any beat period. That is, if the $x_{Ni}(t)$ are sums of terms that vary harmonically, with various discrete frequencies (not necessarily commensurate), then the average is over a time longer than the inverse of the smallest frequency difference. A derivation of Eq. (2.3.3) is given in the appendix at the end of this section.

For a binary star, with the coordinates of the two stars given by Eq. (2.1.1), the mass tensor (2.3.4) is

$$Q_{ij}(t) = m_1 x_{1i}(t) x_{1j}(t) + m_2 x_{2i}(t) x_{2j}(t) = \mu\, r_i(t) r_j(t), \tag{2.3.5}$$

where $\mu \equiv m_1 m_2/(m_1+m_2)$ is the reduced mass, and $\mathbf{r}$ is the separation vector. It follows that the average power (2.3.3) emitted in gravitational radiation is

$$\langle P \rangle = \frac{G\mu^2}{5} \Big\langle 2\dddot{\mathbf{r}}^2 r^2 + 12(\dddot{\mathbf{r}} \cdot \dot{\mathbf{r}})(\dot{\mathbf{r}} \cdot \mathbf{r}) + 12(\dddot{\mathbf{r}} \cdot \dot{\mathbf{r}})(\ddot{\mathbf{r}} \cdot \mathbf{r})$$
$$+ 2(\dddot{\mathbf{r}} \cdot \mathbf{r})^2 + 18 \ddot{\mathbf{r}}^2 \cdot \dot{\mathbf{r}}^2 + 18(\ddot{\mathbf{r}} \cdot \dot{\mathbf{r}})^2 \Big\rangle. \tag{2.3.6}$$

For instance, in the case of zero eccentricity, since r is here constant we have

$$\mathbf{r} = r\hat{r}, \quad \dot{\mathbf{r}} = r\Omega\hat{\varphi}, \quad \ddot{\mathbf{r}} = -r\Omega^2\hat{r}, \quad \dddot{\mathbf{r}} = -r\Omega^3\hat{\varphi},$$

---

[10] J. Larmor, *Phil. Mag.* Series 5. **44**, 503 (1897).

where $\hat{r}$ and $\hat{\varphi}$ are unit vectors in the direction of increasing r and φ, and $\Omega = \sqrt{MG/r^3}$ is the orbital angular velocity. In this case the second, fourth, and sixth terms in Eq. (2.3.6) vanish, while the others add up to

$$\langle P \rangle = \frac{32G\mu^2}{5c^5} \left(\frac{MG}{r^3}\right)^3 r^4.$$

For general ellipticity, we get a factor $\sqrt{MG/a^3}$ for every time derivative, a factor a for every $\mathbf{r}$ or its time derivative, and a factor of a more-or-less complicated function $f(e)$ of ellipticity, so

$$-\left\langle\frac{dE}{dt}\right\rangle = \langle P \rangle = \frac{32G\mu^2}{5c^5} \left(\frac{MG}{a^3}\right)^3 a^4 f(e). \qquad (2.3.7)$$

A detailed calculation[11] of the average (2.3.6) gives

$$f(e) = (1-e^2)^{-7/2} \left[1 + \frac{73}{24}e^2 + \frac{37}{96}e^4\right]. \qquad (2.3.8)$$

Because $f(0) = 1$, this agrees with the result given above for circular orbits, where $r = a$. Similarly, gravitational waves carry away the angular momentum component $J_\perp$ normal to the orbit, at a rate[12]

$$-\left\langle\frac{dJ_\perp}{dt}\right\rangle = \frac{32G^{7/2}\mu^2 M^{5/2}}{5c^5 a^{7/2}(1-e^2)^2} \left(1 + \frac{3e^2}{8}\right). \qquad (2.3.9)$$

Aside from constants, according to Eq. (2.1.7) E depends only on a while $J_\perp$ depends only on L, so from Eqs. (2.3.7)–(2.3.9) we can easily find the rate of decrease of a and L. Since $L = a(1-e^2)$, one can then also find the rate of change of the ellipticity e. According to Peters,[13]

$$\left\langle\frac{de}{dt}\right\rangle = -e \times \frac{304G^3\mu M^2}{15c^5 a^4(1-e^2)^{5/2}} \left(1 + \frac{121}{304}e^2\right). \qquad (2.3.10)$$

The important thing is that the ellipticity decreases, at a rate that accelerates as a decreases, so binary orbits eventually become circular. This is the case for the coalescing binaries discussed in the next section. But, by this standard, the Hulse–Taylor binary pulsar, which has $e \simeq 0.617$, is clearly quite young.

The loss of energy of the binary star is observed as a decrease of its orbital period. Eliminating the semi-major axis a in Eqs. (2.1.6) and (2.1.7) gives the period T in terms of the orbital energy E:

$$T = MG(-2E/\mu)^{-3/2}.$$

---

[11] P. C. Peters and J. Mathews, *Phys. Rev.* **131**, 435 (1963).
[12] P. C. Peters, *Phys. Rev.* **136**, B1224 (1964).
[13] Peters, *op. cit.*

Thus, if E decreases on average at a rate $\langle P \rangle$, the period decreases at an average fractional rate

$$\dot{T}/T = -\frac{3}{2}\langle P \rangle/(-E). \tag{2.3.11}$$

Inserting Eqs. (2.3.7) and (2.1.7), with a given in terms of T by Eq. (2.1.6), we find

$$\dot{T} = -\frac{96\mu G}{5c^5}(2\pi)^{8/3} T^{-5/3} (MG)^{2/3} f(e). \tag{2.3.12}$$

The characteristic sign of the decrease of energy via gravitational radiation is a rate of decrease of the period T that increases as $T^{-5/3}$. For old binaries, with $e = 0$, we have $f(e) = 1$, and Eq. (2.3.7) shows that the quantity $T^{5/3} dT/dt$ depends only on a single mass parameter, known as *the chirp mass*:

$$T^{5/3}\dot{T} = -\frac{96G}{5c^5}(2\pi)^{8/3} m_{\text{chirp}}^5 /3G^{2/3}, \tag{2.3.13}$$

where

$$m_{\text{chirp}} \equiv \mu^{3/5} M^{2/5} = (m_1 m_2)^{3/5}(m_1 + m_2)^{-1/5}.$$

To confirm that the observed rate of decrease of the orbital period of a binary pulsar is really due to the emission of gravitational radiation, we need to know μ, M, and e as well as T and $\dot{T}$. Fortunately, as we have seen, all these parameters are known for the binary pulsar discovered in 1975 by Hulse and Taylor. Using the values found in 2010, Weisberg, Nice, and Taylor calculated a theoretical rate of period decrease $\dot{T} = -2.402531(14) \times 10^{-12}$. Using Eq. (2.3.12) and the same parameters, but with more rounding off, I find $\dot{T} = -2.402 \times 10^{-12}$. Either way, this is in good agreement with the "observed" value $\dot{T} = -2.423(1) \times 10^{-12}$, obtained as a fit to the data. There seems to be no doubt that the decrease of the orbital period is caused by the emission of gravitational radiation, in accordance with the prediction of general relativity.

From the observed orbital period T_0 and its observed rate of decrease $\dot{T}_0$ in a binary star at the present time $t = t_0$, we can calculate the time t_1 when gravitational radiation will bring about the vanishing of the period and the coalescence of the binary's two stars. Since according to Eq. (2.3.12) $\dot{T}$ is proportional to $T^{-5/3} f(e)$, we have

$$t_1 - t_0 = \int_{T_0}^{0} \frac{dT}{\dot{T}(T)} = \int_{T_0}^{0} \frac{f(e_0)\, dT}{\dot{T}_0 (T/T_0)^{-5/3} f(e)} > \frac{3 T_0}{8 |\dot{T}_0|}, \tag{2.3.14}$$

with the final inequality following from the decrease of $f(e)$ during the relatively brief final period when the orbit evolves rapidly toward smaller ellipticity. Using the values $T_0 = 0.323$ days and $\dot{T}_0 = -2.297 \times 10^{-12}$

for the Hulse–Taylor binary pulsar, we see that if no other processes speed up the coalescence, then this binary will coalesce in somewhat more than 144 million years.

However long it takes, the coalescence of a binary of neutron stars or black holes will produce interesting effects. Powerful sources of gamma rays, known as Gamma Ray Bursts (GRBs), had been known since 1967, when they were observed by a US satellite designed to detect Soviet nuclear weapons tests. GRBs were generally supposed to be produced in supernovae, in which a young massive star collapses, but about a third, the short-duration GRBs, were not associated with supernovae or with star-forming regions in which supernovae are expected to occur, and so it seemed plausible that they were instead produced in the coalescence of compact bodies, such as neutron stars and/or black holes. Li and Paczyński[14] noted in 1998 that the coalescence of a pair of neutron stars or of a neutron star and a black hole would eject a small fraction of highly radioactive neutron-rich matter, producing an impressive burst of electromagnetic radiation. In 2010 Metzger et al.[15] estimated that for a day or so the luminosity of such ejecta would be about a thousand times that of typical novae, and therefore called these events *kilonovae*. In 2013 Tanvir et al.[16] reported evidence for such a kilonova, the short gamma ray burst SCRB 130603B and its longer-wavelength afterglow, supporting the hypotheses that the coalescence of compact objects account both for observed short-duration gamma ray bursts and for much of the production of heavy elements. As we will see in the next section, this was confirmed by the observation of a short-duration GRB and then its afterglow shortly after a detection in 2017 of the gravitational wave signal GW170817.[17]

Just before coalescence, a binary of neutron stars and/or black holes will emit a fair fraction of the energy in its rest mass as gravitational radiation in its final revolutions. Using Eq. (2.1.6), the radiated gravitational wave power (2.3.7) can be expressed in terms of the orbital period T instead of the semi-major axis a:

$$\langle P \rangle = \frac{32 G \mu^2 (MG)^{4/3} f(e)}{5 c^5} \left(\frac{T}{2\pi} \right)^{-10/3}.$$

Also, using (2.3.12), we have

$$dt = \frac{5 c^5 T^{5/3} \, dT}{96 \mu G (2\pi)^{8/3} (MG)^{2/3} f(e)},$$

---

[14] L.-X. Li and B. Paczyński, *Astrophys. J.* **507**, L59 (1998).
[15] B. D. Metzger et al., *Mon. Not. Roy. Astron. Soc.* **406**, 2650 (2010).
[16] N. R. Tanvir et al., *Nature* **500**, 547 (2013).
[17] This gamma ray signal was much less luminous than is typical for short-duration gamma ray bursts, so its identification remains somewhat in doubt.

so the total energy radiated when the period drops from a value T_0 at an initial time t_0 to a much smaller value T_1 at a later time t_1 is

$$-\Delta E = \int_{t_1}^{t_0} \langle P \rangle \, dt \simeq \int_{T_1}^{T_0} \frac{(2\pi)^{2/3} \mu (MG)^{2/3}}{3T^{5/3}} \, dT$$

$$\simeq \frac{(2\pi)^{2/3} \mu (MG)^{2/3}}{2T_1^{2/3}}. \tag{2.3.15}$$

For neutron stars coalescence occurs at around the time that the decrease in the size of the orbit would bring the stars into contact. At late times the eccentricity of the orbit will be very small, so coalescence occurs when a reaches a value equal to the sum of the radii of the neutron stars. For massive neutron stars this is of order MG/c^2, the minimum period T_1 is thus of order $2\pi MG/c^3$, and the total energy radiated is of order μc^2.

We can be a bit more precise about black holes. As shown in the appendix to Section 4.5, the minimum radius of a stable circular orbit of a test body revolving about a black hole of mass M has radius $R = 6MG/c^2$. Using this as a rough estimate for a binary black hole of total mass M, we can use Eq. (2.1.6) to guess that a black hole binary continues in orbit until its period reaches a minimum value $T_1 = 2\pi 6^{3/2} MG/c^3$, after which the black holes plunge together and coalesce. The total power (2.3.15) radiated until this plunge is then roughly of order $\mu c^2/12$. Of course, such estimates are subject to large relativistic corrections.

Since the discovery of PSR 1913+16, dozens of pulsars have been found in binary systems, coupled with ordinary stars, white dwarfs, and other neutron stars. (There is even a binary PSR J0737-3038 composed of two pulsars, discovered in 2003 at the Parkes Observatory in Australia.) Although all these binaries are slowing down through the emission of gravitational radiation, so far unfortunately it has not been possible to detect the gravitational radiation from any binary pulsar. The direct observation of gravitational radiation had to wait for the detection of the enormous energy radiated by coalescing binaries, to be discussed in the next section.

Appendix: Review of Gravitational Radiation

This book is not a treatise on general relativity, but in view of the increasing importance in astronomy of gravitational radiation, it seems appropriate to provide a brief survey here of the theory of this radiation. Derivations are generally abbreviated or skipped.[18]

---

[18] The full derivations can be found in many books, including Chapter 10 of S. Weinberg, *Gravitation and Cosmology* (Wiley, New York, 1972). Derivations that were not given in this reference, including the derivation of the quadrupole approximation for emitted power in the form (2.3.A36), are given here.

2.3 Gravitational Wave Emission: Binary Pulsars

For a weak gravitational field, the metric is

$$g_{\mu\nu}(x) = \eta_{\mu\nu} + h_{\mu\nu}(x), \quad |h_{\mu\nu}(x)| \ll 1, \tag{2.3.A1}$$

where μ and ν are spacetime indices, running over the values 1, 2, 3, 0, with $x^0 = t$, and the non-zero components of $\eta_{\mu\nu}$ are $\eta_{11} = \eta_{22} = \eta_{33} = 1$ and $\eta_{00} = -1$.[19] It is always possible to adopt what is called a harmonic coordinate system, which in the weak field case satisfies the condition

$$\frac{\partial}{\partial x^\mu} h^\mu{}_\nu(x) = \frac{1}{2} \frac{\partial}{\partial x^\nu} h^\mu{}_\mu(x). \tag{2.3.A2}$$

Here and elsewhere in using the weak field approximation, indices are raised and lowered with $\eta_{\mu\nu}$, and repeated indices are summed.

The Einstein field equations for weak gravitational fields in harmonic coordinates read

$$\Box h_{\mu\nu}(x) = -16\pi G S_{\mu\nu}(x), \tag{2.3.A3}$$

where $\Box$ is the d'Alembertian $\Box \equiv \eta^{\mu\nu} \partial^2/\partial x^\mu \partial x^\nu$, and $S_{\mu\nu}(x)$ is related to the energy-momentum tensor $T_{\mu\nu}(x)$ by

$$S_{\mu\nu}(x) \equiv T_{\mu\nu}(x) - \frac{1}{2} \eta_{\mu\nu} T^\lambda{}_\lambda(x). \tag{2.3.A4}$$

Let us first consider the solution for empty space, in which case the field equation is

$$\Box h_{\mu\nu}(x) = 0. \tag{2.3.A5}$$

We can show that by a suitable choice of coordinates that preserves the harmonic coordinate condition (2.3.A2) we can always eliminate the time components of $h_{\mu\nu}$, so that

$$h_{i0} = h_{0i} = h_{00} = 0, \tag{2.3.A6}$$

(with i and j running over the space coordinate labels 1, 2, 3), and it then follows from Eq. (2.3.A2) that, apart from possible time-independent terms,

$$h_{ii} = 0, \quad \frac{\partial h_{ij}}{\partial x_j} = 0. \tag{2.3.A7}$$

[Here is the proof. Under a general coordinate transformation $x^\mu \mapsto x^\mu + \epsilon^\mu(x)$, with $\epsilon^\mu(x)$ small (of the same order as $h_{\mu\nu}(x)$), we have

$$h_{\mu\nu} \mapsto h_{\mu\nu} + \frac{\partial \epsilon_\mu}{\partial x^\nu} + \frac{\partial \epsilon_\nu}{\partial x^\mu}.$$

We preserve the condition (2.3.A2) by requiring that $\Box \epsilon_\mu = 0$. We can make h_{00} vanish by taking ϵ_0 to satisfy $\partial \epsilon_0/\partial x^0 = -h_{00}/2$, and we can make h_{0i} vanish by taking ϵ_i to satisfy $\partial \epsilon_i/\partial x^0 = -h_{0i} - \partial \epsilon_0/\partial x^i$. Setting $\nu = 0$ in

---
[19] Until the end of this appendix, the speed of light is set equal to unity.

Eq. (2.3.A2) then gives $0 = \partial h_{ii}/\partial x^0$, so, apart from time-independent terms, $h_{ii} = 0$. It then follows from Eq. (2.3.A2) that also $\partial h_{ij}/\partial x^j = 0$.]

If we further assume that $h_{\mu\nu}$ depends only on x_3 and time, as for a wave traveling in the ± 3-direction, then the vanishing of $\partial h_{ij}/\partial x_j$ tells us that $\partial h_{i3}/\partial x_3 = 0$, so, apart from terms that do not depend on $\mathbf{x}$, we also have

$$h_{i3} = h_{3i} = 0. \tag{2.3.A8}$$

This leaves $h_{\mu\nu}$ with only two independent non-zero components,

$$h_+ \equiv h_{11} = -h_{22}, \quad h_\times \equiv h_{12} = h_{21}. \tag{2.3.A9}$$

Such a gravitational wave is said to be in *transverse-traceless gauge*. By considering how these components transform under rotations around the 3-axis, one can conclude that the components for gravitons with helicity $\pm 2\hbar$ are $h_+ \mp i h_\times$.

It should be noted that for a wave that propagates in the 3-direction, for which $h_{\mu\nu}$ does not depend on x^1 or x^2, the coordinate shifts ϵ^μ that are used to put $h_{\mu\nu}$ into transverse-traceless gauge do not depend on x^1 or x^2, so the transformation $h_{\mu\nu} \mapsto h_{\mu\nu} + \partial \epsilon_\mu/\partial x^\nu + \partial \epsilon_\nu/\partial x^\mu$ has no effect on the components (2.3.A9). Thus, to find the components (2.3.A9) in transverse-traceless gauge, there is no need actually to transform to this gauge; it is only necessary to inspect the space–space components h_{ij} with i and j taking the values 1 and/or 2.

The general solution of Eqs. (2.3.A5) and (2.3.A2) is a linear superposition of terms of the form

$$h_{\mu\nu}(x) = e_{\mu\nu} e^{i k_\lambda x^\lambda} + e^*_{\mu\nu} e^{-i k_\lambda x^\lambda}, \tag{2.3.A10}$$

where the k^λ are real constant wave vectors satisfying

$$k^\lambda k_\lambda = 0, \tag{2.3.A11}$$

and the $e_{\mu\nu} = e_{\nu\mu}$ are complex constant polarization tensors satisfying

$$k_\mu e^\mu{}_\nu = \frac{1}{2} k_\nu e^\mu{}_\mu. \tag{2.3.A12}$$

In transverse-traceless gauge, the polarization tensor satisfies the further conditions

$$e_{0i} = e_{i0} = e_{00} = 0, \quad e^i{}_i = 0, \quad k_i e^{ij} = 0, \tag{2.3.A13}$$

where as usual i and j run over the values 1, 2, 3. This leaves only two independent components, which when k^i is in the 3-direction are

$$e_{11} = -e_{22} \equiv e_+, \quad e_{12} = e_{21} \equiv e_\times. \tag{2.3.A14}$$

Let us now return to the field equation (2.3.A3) in the presence of a non-zero source. The general solution is

$$h_{\mu\nu}(\mathbf{x}, t) = 4G \int d^3 x' \left(\frac{S_{\mu\nu}(\mathbf{x}', t - |\mathbf{x} - \mathbf{x}'|)}{|\mathbf{x} - \mathbf{x}'|} \right), \tag{2.3.A15}$$

plus any linear combination of free-field solutions of form (2.3.A10). Suppose the source is a sum of simple harmonic terms with various angular frequencies ω (not necessarily commensurate)[20]

$$T_{\mu\nu}(\mathbf{x}, t) = \sum_{\omega} \left[T_{\mu\nu}(\mathbf{x}, \omega) \exp(-i\omega t) + T^*_{\mu\nu}(\mathbf{x}, \omega) \exp(i\omega t) \right]. \quad (2.3.\text{A}16)$$

Then in the wave zone, where $\omega |\mathbf{x} - \mathbf{x}'| \gg 1$ for all the frequencies ω in Eq. (2.3.A16) and for all $\mathbf{x}'$ in the source (that is, for which $T_{\mu\nu}(\mathbf{x}', \omega)$ is non-zero), the solution (2.3.A15) becomes a sum of plane waves like (2.3.A10) for the various frequencies ω:

$$h_{\mu\nu}(\mathbf{x}, t) = \sum_{\omega} \left[e_{\mu\nu}(\omega, \mathbf{x}) e^{i\mathbf{k}(\omega,\hat{x})\cdot\mathbf{x}} e^{-i\omega t} + e^*_{\mu\nu}(\omega, \mathbf{x}) e^{-i\mathbf{k}(\omega,\hat{x})\cdot\mathbf{x}} e^{i\omega t} \right],$$

(2.3.A17)

with

$$\mathbf{k}(\omega, \hat{x}) = \omega \hat{x}, \qquad \hat{x} \equiv \mathbf{x}/|\mathbf{x}|, \quad (2.3.\text{A}18)$$

$$e_{\mu\nu}(\omega, \mathbf{x}) = \frac{4G}{|\mathbf{x}|} \left[T_{\mu\nu}\big(\mathbf{k}(\omega, \hat{x}), \omega\big) - \frac{1}{2}\eta_{\mu\nu} T^{\lambda}{}_{\lambda}\big(\mathbf{k}(\omega, \hat{x}), \omega\big) \right], \quad (2.3.\text{A}19)$$

and

$$T_{\mu\nu}(\mathbf{k}, \omega) \equiv \int d^3x' \, T_{\mu\nu}(\mathbf{x}', \omega) \exp\left(-i\mathbf{k}\cdot\mathbf{x}'\right). \quad (2.3.\text{A}20)$$

Although both $\mathbf{k}(\omega, \hat{x})$ and $e_{\mu\nu}(\omega, \mathbf{x})$ depend on $\mathbf{x}$, in the wave zone they change by negligible amounts when $\mathbf{x}$ changes by amounts of the order of the wavelength $1/|\omega|$.

Energy-momentum conservation guarantees

$$k_i T_{i\nu}(\mathbf{k}, \omega) = \omega T_{0\nu}(\mathbf{k}, \omega), \quad (2.3.\text{A}21)$$

from which it follows that (2.3.A19) satisfies the harmonic coordinate condition (2.3.A12), and with which we can express all components of $T_{\mu\nu}$ in terms of the space–space components:

$$T_{0j}(\mathbf{k}, \omega) = T_{j0}(\mathbf{k}, \omega) = k_i T_{ij}(\mathbf{k}, \omega)/\omega, \quad (2.3.\text{A}22)$$

$$T_{00}(\mathbf{k}, \omega) = k_i T_{i0}(\mathbf{k}, \omega)/\omega = k_i k_j T_{ij}(\mathbf{k}, \omega)/\omega^2. \quad (2.3.\text{A}23)$$

In general (2.3.A19) does not satisfy the transverse-traceless conditions (2.3.A13) without a further coordinate transformation, but as previously mentioned this transformation does not change the components e_+ and $e_\times$, so these components can be read off from Eq. (2.3.A19) without needing

---

[20] In this section we commit the notational sin of using the same symbol for functions and their Fourier transforms or Fourier components, leaving it to the displayed arguments to indicate which is intended.

to perform this transformation. In particular, for **x** and hence $\mathbf{k}(\omega, \hat{x})$ in the 3-direction,

$$e_\times(\omega, \mathbf{x}) = \frac{4G}{|\mathbf{x}|} T_{12}(\mathbf{k}(\omega, \hat{x}), \omega),$$

$$e_+(\omega, \mathbf{x}) = \frac{2G}{|\mathbf{x}|} \left[T_{11}(\mathbf{k}(\omega, \mathbf{x}), \omega) - T_{22}(\mathbf{k}(\omega, \mathbf{x}), \omega) \right]. \quad (2.3.\text{A}24)$$

(In deriving the formula for e_+, we use Eq. (2.3.A23), which for $\mathbf{k}(\omega, \hat{x})$ in the 3-direction gives $T_{00}(\mathbf{k}(\omega, \hat{x}), \omega) = T_{33}(\mathbf{k}(\omega, \hat{x}), \omega)$.)

We evidently only need the spatial components of $T_{\mu\nu}(\mathbf{k}, \omega)$. These are greatly simplified in what is known as the quadrupole approximation. The typical speed of matter in the source is $\omega|\mathbf{x}'|$, where $\mathbf{x}'$ is the coordinate separation from a point (such as the center of mass) relative to which velocities are measured. Hence, if the matter of the source is moving non-relativistically, we have $\omega|\mathbf{x}'| \ll 1$. Using Eq. (2.3.A23), expanding Eq. (2.3.A20) for $T_{00}(\mathbf{k}, \omega)$ to second order in $\mathbf{k}$, and matching coefficients of $k_i k_j$, we have

$$T_{ij}(\mathbf{k}, \omega) \simeq -\frac{\omega^2}{2} Q_{ij}(\omega), \quad (2.3.\text{A}25)$$

where Q_{ij} is the mass tensor

$$Q_{ij}(\omega) = \int d^3 x' \, x_i' x_j' T_{00}(\mathbf{x}', \omega). \quad (2.3.\text{A}26)$$

Using Eqs. (2.3.A16), (2.3.A18), (2.3.A24), (2.3.A25), and (2.3.A26) in Eq. (2.3.A17), we find the components of the time-dependent gravitational wave amplitude in transverse-traceless gauge for a source at a distance d in the 3-direction

$$h_\times(\mathbf{x}, t) = \frac{2G}{d} \ddot{Q}_{12}(t-d), \quad h_+(\mathbf{x}, t) = \frac{G}{d} \left[\ddot{Q}_{11}(t-d) - \ddot{Q}_{22}(t-d) \right], \quad (2.3.\text{A}27)$$

where $Q_{ij}(t)$ are the components of the time-dependent mass tensor

$$Q_{ij}(t) = \int d^3 x \, x_i x_j T_{00}(\mathbf{x}, t) = \sum_\omega \left[e^{-i\omega t} Q_{ij}(\omega) + e^{i\omega t} Q_{ij}^*(\omega) \right]. \quad (2.3.\text{A}28)$$

Although contemporary observations of gravitational radiation rely on the detection of spatial distortion rather than of energy flux, it is important to know the energy radiated gravitationally by sources such as binary stars, in order to learn the effect of this radiation on the sources. The energy and momentum in gravitational waves can be obtained from a pseudo-tensor $t_{\mu\nu}$, defined by moving the non-linear terms on the left-hand side $R_{\mu\nu} - g_{\mu\nu} R^\lambda{}_\lambda / 2$ of the Einstein field equations to the right-hand side, where they act as a source to

2.3 Gravitational Wave Emission: Binary Pulsars

the linearized Einstein equations. To second order in the perturbation $h_{\mu\nu}$ to the Minkowski metric,

$$t_{\mu\nu} = \frac{1}{8\pi G}\left[R^{(2)}_{\mu\nu} - \frac{1}{2}\eta_{\mu\nu}\eta^{\rho\sigma}R^{(2)}_{\rho\sigma}\right], \qquad (2.3.\text{A}29)$$

where $R^{(2)}_{\mu\nu}$ is the part of the Ricci tensor $R_{\mu\nu}$ of second order in the perturbation to the Minkowski metric. ($t_{\mu\nu}$ is a pseudo-tensor, in the sense that it transforms as a tensor under Lorentz transformation, but not under general coordinate transformations.) Since t^{i0} is the energy flux vector, the energy per time radiated in a solid angle $d\Omega$ around a direction $\hat{x}$ is $dP = |\mathbf{x}|^2 \hat{x}_i t^{i0}\, d\Omega$. Evaluating t^{i0} in the wave zone using Eq. (2.3.A17) and averaging over a time that is long compared with the longest beat frequency (the inverse of the smallest difference between frequencies ω), the average radiated power per solid angle is

$$\left\langle \frac{dP(\mathbf{x})}{d\Omega}\right\rangle = \frac{|\mathbf{x}|^2}{16\pi G}\sum_{\omega}\omega^2\left[e^{\mu\nu*}(\omega,\mathbf{x})e_{\mu\nu}(\omega,\mathbf{x}) - \frac{1}{2}\left|e^{\mu}{}_{\mu}(\omega,\mathbf{x})\right|^2\right]. \qquad (2.3.\text{A}30)$$

Using Eq. (2.3.A19), we see that this depends only on the direction of $\mathbf{x}$:

$$\left\langle \frac{dP(\hat{x})}{d\Omega}\right\rangle = \frac{G}{\pi}\sum_{\omega}\omega^2\left[T^{\mu\nu*}(\mathbf{k}(\hat{x},\omega),\omega)T_{\mu\nu}(\mathbf{k}(\hat{x},\omega),\omega)\right.$$
$$\left. -\frac{1}{2}\left|T^{\mu}{}_{\mu}(\mathbf{k}(\hat{x},\omega),\omega)\right|^2\right]. \qquad (2.3.\text{A}31)$$

Equations (2.3.A22) and (2.3.A23) allow us to write this in terms of the space components T_{ij}:

$$\left\langle \frac{dP(\hat{x})}{d\Omega}\right\rangle = \frac{G}{\pi}\sum_{\omega}\omega^2 T^{ij*}(\mathbf{k}(\hat{x},\omega),\omega)T^{kl}(\mathbf{k}(\hat{x},\omega),\omega)$$
$$\times \left(\delta_{il}\delta_{jm} - \frac{1}{2}\delta_{ij}\delta_{lm} - 2\hat{x}_j\hat{x}_m\delta_{il} + \frac{1}{2}\hat{x}_l\hat{x}_m\delta_{ij}\right.$$
$$\left. + \frac{1}{2}\hat{x}_i\hat{x}_j\delta_{lm} + \frac{1}{2}\hat{x}_i\hat{x}_j\hat{x}_k\hat{x}_l\right). \qquad (2.3.\text{A}32)$$

The angular dependence of the emitted power becomes much more explicit in the quadrupole approximation, in which $T_{ij}(\mathbf{k},\omega)$ is given by a tensor (2.3.A25) independent of $\mathbf{k}$. It is then easy to integrate over solid angle, which gives the total average emitted power

$$\langle P \rangle = \frac{2G}{5}\sum_{\omega}\omega^6\left[Q^*_{ij}(\omega)Q_{ij}(\omega) - \frac{1}{3}|Q_{ii}(\omega)|^2\right]. \qquad (2.3.\text{A}33)$$

Equation (2.3.A33) is the classic formula for emitted gravitational radiation power in the quadrupole approximation. Using it is convenient for a source that has a simple harmonic time-dependence, with only a single frequency ω, such as a binary star with a circular orbit. But for sources whose time-dependence is a superposition of many frequencies, such as a binary star with an elliptical orbit, it is much more convenient to write the emitted power in terms of the actual time-dependence of the mass tensor (2.3.A28).

We want to show that the power (2.3.A33) is given by the time-average of products of third time derivatives of the mass tensor (2.3.A28). Consider the average

$$\left\langle \frac{d^3}{dt^3} Q^*_{ij}(t) \frac{d^3}{dt^3} Q_{kl}(t) \right\rangle = \sum_{\omega,\omega'} \omega^3 \omega'^3 \left\langle \left[e^{+i\omega t} Q^*_{ij}(\omega) - e^{-i\omega t} Q_{ij}(\omega) \right] \right. \\ \left. \times \left[e^{-i\omega t} Q_{kl}(\omega') - e^{+i\omega t} Q^*_{kl}(\omega') \right] \right\rangle. \quad (2.3.A34)$$

The averaging over times long compared with the longest of the beat periods $1/|\omega - \omega'|$ kills all terms with $\omega \neq \omega'$, and all terms with time-dependence $\exp(\pm 2i\omega t)$, so Eq. (2.3.A34) gives

$$\left\langle \dddot{Q}^*_{ij}(t) \dddot{Q}_{kl}(t) \right\rangle = \sum_{\omega} \omega^6 \left[Q^*_{ij}(\omega) Q_{kl}(\omega) + Q_{ij}(\omega) Q^*_{kl}(\omega) \right]. \quad (2.3.A35)$$

Comparing this with Eq. (2.3.A33), we see that the average power emitted in gravitational radiation is

$$\langle P \rangle = \frac{G}{5c^5} \left\langle \dddot{Q}^*_{ij}(t) \dddot{Q}_{ij}(t) - \frac{1}{3} \left| \dddot{Q}_{ii}(t) \right|^2 \right\rangle. \quad (2.3.A36)$$

We have used dimensional analysis to include a factor c^5 in the denominator here, in order to make this formula correct in cgs units, and therefore ready to use for numerical calculation.

2.4 Gravitational Wave Detection: Coalescing Binaries

In the 1960s and 1970s interest in the detection of gravitational waves centered on measurement of the energy they deposit as elastic waves in large metal cylinders. In 1969 Joseph Weber[21] claimed to have detected gravitational waves with such a device, but Weber's cylinder was chiefly sensitive to sources emitting gravitational waves with a frequency that happens to match one of its normal modes, and there was no good candidate for a source with such a frequency, or otherwise strong enough to be observed with Weber's apparatus.

Since then, it has been generally agreed that this supposed detection was in error. Attention has shifted to a different sort of detector, conceived in 1972

---

[21] J. Weber, *Phys. Rev. Lett.* **22**, 1320 (1969); **24**, 276 (1970); **25**, 180 (1970).

by Rainer Weiss,[22] which seeks instead to observe changes in phase of electromagnetic waves in the arms of a laser interferometer. A National Science Foundation project instigated by Weiss with Ronald Drever, Kip Thorne, and others, and then with the supervision of Barry Barish, built such an observatory, the Laser Interferometric Gravitational Wave Observatory (LIGO), with installations at Livingston, Louisiana and Hanford, Washington, each consisting of two 4 kilometer arms at right angles. A European collaboration built a similar interferometer, Virgo, at Cascina in Italy, and plans are in train for installations in other countries. Such detectors are sensitive to waves in a range of frequencies, not just a few normal modes.

LIGO operated from 2002 to 2010, and reported no gravitational wave detection. It was then succeeded by an advanced version.[23] As we shall see, Advanced LIGO soon detected what could only be interpreted as gravitational waves from the coalescence of a binary consisting of a pair of black holes.

For orientation, let's first make a crude estimate of the gravitational field perturbation to be expected far from typical sources. For a monochromatic source that produces gravitational waves with frequency $\omega/2\pi$, Eqs. (2.3.A17)–(2.3.A20) of the appendix to the previous section give a typical value $\overline{h}$ of the components of the perturbation $h_{\mu\nu}$ to $g_{\mu\nu}$ at a distance d from the source as $\overline{h} \approx G\overline{T}/dc^4$, where $\overline{T}$ is a typical value of Fourier components of the energy-momentum tensor, and we now use dimensional analysis to put in powers of c to make expressions valid in cgs units. As shown in Eqs. (2.3.A25) and (2.3.A26), in the quadrupole approximation for a source consisting of small bodies with total mass M, linear extent a, and typical velocities $\overline{v} \approx a\omega \ll c$, we have $\overline{T} \approx M\overline{v}^2$, so

$$\overline{h} \approx MG\overline{v}^2/dc^4. \tag{2.4.1}$$

For the Hulse–Taylor binary pulsar discussed in Section 2.3, $d \simeq 6.4$ kpc $= 2 \times 10^{22}$ cm, $M \simeq 2.8 M_\odot$, $MG/c^2 \simeq 4 \times 10^5$ cm, and $\overline{v}/c \simeq 10^{-3}$, so Eq. (2.4.1) gives $\overline{h} \approx 2 \times 10^{-24}$. As we shall see, it is hopeless for LIGO to detect a gravitational wave this weak with a period as long as the 8 hour period of this binary pulsar.

Other sources are more promising. When a pair of neutron stars in orbit around each other are just about to coalesce, their minimum separation $a(1-e)$ becomes equal to the sum R of their radii. Also, as shown in Eq. (2.3.10), the loss of energy and angular momentum through gravitational radiation will by then generally have reduced e to a small value, so $a \simeq R$. For a pair of neutron stars R will typically be of order 30 km, so if $M \simeq 3M_\odot$ the orbital frequency at coalescence will be

---

[22] Quarterly Report of the Research Laboratory for Electronics, MIT Report No. 105 (1972).
[23] For a detailed description of Advanced LIGO, see J. Aasi *et al.* (LIGO Scientific Collaboration), *Classical & Quantum Gravity* **32**, 074001 (2015).

$$\frac{1}{T} = \frac{1}{2\pi}\sqrt{\frac{MG}{R^3}} \simeq 6 \times 10^2 \text{ Hz}.$$

The typical velocity will be of order $\omega a = 2\pi a/T$, which is then $\bar{v}/c \simeq 0.4$. If the coalescing neutron star binary is at the same distance as the Hulse–Taylor binary pulsar, then Eq. (2.4.1) gives $\bar{h} \approx 3 \times 10^{-19}$, and of course greater or less if the coalescing binary is closer or farther.

On the other hand, the gravitational wave signal from a pair of coalescing black holes is somewhat greater than from a pair of neutron stars of the same mass and distance. This is because the Schwarzschild radius of a black hole is smaller than the radius of a neutron star of the same mass, where the neutron star mass is well below the Landau–Oppenheimer–Volkoff limit. For this reason, coalescence occurs not when the separation equals the sum of their Schwarzschild radii, but earlier, when their orbit becomes unstable. As shown in the appendix to Section 4.5 for the case where one body is much heavier than the other, this happens when the separation is three times the bigger Schwarzschild radius, but this is still smaller than the sum of the radii of neutron stars of the same mass. Still, there is not much difference in the signal received from coalescing black holes or neutron stars at the same mass and distance. As it has turned out, the important difference between mergers of neutron stars and black holes as sources of gravitational radiation is that black hole masses are not limited by an upper bound like the Landau–Oppenheimer–Volkoff bound on the masses of stable neutron stars, and, as we shall see, have been discovered with masses much larger than any stable neutron star.

Whether it is neutron stars or black holes that are coalescing, or one of each, the leading signature of a coalescence is a wave with a period T that according to Eq. (2.3.12) decreases at a rate proportional to $T^{-5/3}$. The solution of the equation $\dot{T} \propto T^{-5/3}$ is

$$T \propto (t - t_1)^{3/8},$$

where t_1 is the time of coalescence. This specific result for the decrease of period applies only within the quadrupole approximation, but in general we expect the period of the emitted gravitational wave to decrease as the binary spirals inward, giving a signal known as a *chirp*.

Let's now look at the geometry of LIGO and similar gravitational wave observatories. A laser, operating at an optical or infrared wavelength, sends a coherent electromagnetic wave into the observatory, say along the 1-axis. The wave strikes a beam splitter, with the effect that a transmitted portion continues along an arm in the 1-direction, while a reflected part is sent along an arm in the orthogonal 2-direction. Both the transmitted wave and the reflected wave strike mirrors, and are reflected back to the beam splitter, where a portion recombines and continues along the negative 2-axis to a detector. In practice, the effective path lengths of the two arms are increased by a factor of order 100 by

2.4 Gravitational Wave Detection: Coalescing Binaries

interposing a partly silvered mirror in each arm near the beam splitter, so that each electromagnetic wave is reflected back and forth many times before it reaches the detector.

Suppose that a weak gravitational wave with metric $g_{\mu\nu} = \eta_{\mu\nu} + h_{\mu\nu}$ enters the observatory. We adopt a coordinate system for which to zeroth order in $h_{\mu\nu}$ the mirrors are at rest. In laser interferometers like LIGO, the mirrors are suspended so that their natural period of oscillation is much longer than the characteristic frequencies of the gravitational waves to which they are sensitive, so that to a good approximation the only horizontal forces to which they are exposed are gravitational. The spacetime coordinate x^μ of each mirror thus satisfies the equation for a freely falling body

$$\frac{d^2 x^\mu}{d\tau^2} + \Gamma^\mu_{\rho\sigma}(x) \frac{dx^\rho}{d\tau} \frac{dx^\sigma}{d\tau} = 0, \qquad (2.4.2)$$

where

$$\Gamma^\mu_{\rho\sigma} = \frac{1}{2} g^{\mu\nu} \left[\frac{g_{\nu\rho}}{\partial x^\sigma} + \frac{g_{\nu\sigma}}{\partial x^\rho} - \frac{g_{\rho\sigma}}{\partial x^\nu} \right] \qquad (2.4.3)$$

and

$$d\tau^2 = -g_{\rho\sigma} dx^\rho dx^\sigma. \qquad (2.4.4)$$

Since $\Gamma^\mu_{\rho\sigma}$ is at least of first order in $h_{\mu\nu}$, to this order on the right-hand side of Eq. (2.4.2) we can take $dx^i/d\tau = 0$ and $dt/d\tau = 1$, and

$$\Gamma^i_{00} = \frac{\partial h_{j0}}{\partial t} - \frac{1}{2} \frac{\partial h_{00}}{\partial x^i}, \qquad \Gamma^0_{00} = -\frac{1}{2} \frac{\partial h_{00}}{\partial t}, \qquad (2.4.5)$$

so, according to Eq. (2.4.2), the change δx^μ in the coordinates due to the gravitational wave satisfies

$$\frac{\partial^2 \delta x^i}{\partial t^2} = -\frac{\partial h_{j0}}{\partial t} + \frac{1}{2} \frac{\partial h_{00}}{\partial x^i}, \qquad \frac{\partial^2 \delta x^0}{\partial t^2} = \frac{1}{2} \frac{\partial h_{00}}{\partial t}. \qquad (2.4.6)$$

(We are now writing the time derivatives on the left-hand side of these equations of motion as partial derivatives, because δx^μ depends not only on time, but also on the zeroth-order position **x**. As usual, i, j, etc. run over the values 1, 2, 3, and repeated indices are summed over these values.) It is therefore very convenient to adopt a generalized transverse-traceless gauge, in which

$$h_{0i}(x) = h_{i0}(x) = h_{00}(x) = 0, \qquad (2.4.7)$$

and, as a consequence of Eq. (2.3.A2),

$$\partial_i h_{ij} = 0, \qquad h_{ii} = 0, \qquad (2.4.8)$$

so in this gauge there is no first-order effect of the gravitational wave on the coordinates of the mirrors. The reader can check that this choice of gauge is accomplished by the coordinate transformation $x^\mu \to x^\mu - \delta x^\mu$, with δx^μ satisfying Eq. (2.4.6). (Equations (2.4.7) and (2.4.8) are satisfied after a coordinate transformation for the plane waves discussed in Section 2.3, but we are not now assuming that the wave is simple-harmonic or traveling in the 3-direction.)

We now wish to consider the effect of the gravitational wave on an electromagnetic wave traveling between two mirrors separated by a fixed coordinate distance L. To first order in h, we have $g^{\mu\nu} = \eta^{\mu\nu} - h^{\mu\nu}$, so any individual component $\mathcal{E}(x)$ of an electromagnetic wave will satisfy the equation

$$0 = \left[\eta^{\mu\nu} - h^{\mu\nu}(x)\right] \partial_\mu \partial_\nu \mathcal{E}(x). \tag{2.4.9}$$

If (as is usual) the gravitational perturbation varies little over the arms of the interferometer and the travel time of the electromagnetic wave, we can use a plane wave solution:

$$\mathcal{E}(x) \propto \exp\left(-i\omega t + iq\hat{n}\cdot\mathbf{x}\right), \tag{2.4.10}$$

where $\hat{n}$ is the direction of propagation of the electromagnetic wave and $\omega/2\pi$ is its frequency. (There is of course no necessary relation between ω and the frequency 2Ω of the gravitational wave; in fact, in all relevant cases, $\omega \gg \Omega$.) In transverse-traceless gauge the wave number q therefore satisfies

$$q^2[1 - h_{ij}\hat{n}_i\hat{n}_j] - \omega^2 = 0,$$

which has the first-order solution

$$q = \omega\left[1 + \frac{1}{2}h_{ij}\hat{n}_i\hat{n}_j\right]. \tag{2.4.11}$$

Thus a gravitational wave produces a change in phase of the electromagnetic wave after traveling an effective distance L back and forth along the direction $\hat{n}$ (perhaps bouncing back and forth many times) equal to

$$\Delta\Phi = \frac{1}{2}h_{ij}\hat{n}_i\hat{n}_j\omega L. \tag{2.4.12}$$

We see that when the electromagnetic wave is recombined after traveling effective lengths L_1 and L_2 back and forth in the interferometer arms along the 1- and 2-directions, the wave will take the form

$$\exp(-i\omega t)\left[A_1 \exp\left(i\omega L_1[1 + h_{11}/2]\right) + A_2 \exp\left(i\omega L_2[1 + h_{22}/2]\right)\right] + \text{c.c.}, \tag{2.4.13}$$

where A_1 and A_2 are amplitudes that reflect the effects of reflection and transmission along the two interferometer arms, and c.c. denotes the complex conjugate. Since it is harder to observe a small change in the electromagnetic wave intensity caused by a gravitational wave than to observe a weak electromagnetic wave where previously there were none, LIGO was designed to arrange for

2.4 Gravitational Wave Detection: Coalescing Binaries

destructive interference in the absence of the gravitational wave.[24] The system is adjusted so that to the greatest extent possible $|A_1| = |A_2|$ and $L_1 = L_2 \equiv L$, and then finer adjustments are made so that in the absence of gravitational waves nearly complete destructive interference is observed in the detector. We then have

$$A_1 \exp[i\omega L_1] = -A_2 \exp[i\omega L_2] \equiv A. \tag{2.4.14}$$

In the presence of a gravitational wave the destructive interference will not be complete, and to first order in h the amplitude will be

$$i A \exp(i\omega t) L\omega (h_{11} - h_{22})/2 + \text{c.c.}. \tag{2.4.15}$$

That is, if N photons would have been received at the detector during a time $2\pi/\omega$ if one arm of the interferometer were blocked, so that there is no destructive interference, then the number actually received in the presence of a gravitational wave will be

$$N \left| L\omega (h_{11} - h_{22})/2 \right|^2, \tag{2.4.16}$$

with the components h_{ij} evaluated at the position of the interferometer (which we are assuming is small compared with the wavelength of the gravitational wave) and at the time of the measurement.

The detection of a gravitational wave of frequency $2\Omega/2\pi$ is impeded by two main types of background noise, with very different dependence on Ω.

One noise type is ordinary seismic noise, which increases with decreasing Ω. This is minimized by hanging the interferometers' mirrors on pendula with very low response frequencies, and resting these pendula on seismic isolation tables. Seismic noise can be further suppressed by recording only detections at the two LIGO sites in Louisiana and Washington that arrive in coincidence, within the ≈ 0.01 seconds travel time of a gravitational wave between the two sites. Even so, seismic noise is a serious problem for low-frequency sources, in particular making it impossible to detect a source like the Hulse–Taylor binary pulsar with a period of 8 hours.

The other chief source of noise is "shot noise," arising from the limited number of photons N delivered in a finite time by an electromagnetic wave. This number is subject to quantum fluctuations of order $\sqrt{N}$, and hence fractional fluctuations of order $1/\sqrt{N}$. Since the number of photons arriving during a time of order $1/\Omega$ is proportional to $1/\Omega$, the effect of shot noise increases with increasing Ω. To mitigate the effects of shot noise LIGO uses a powerful laser, but shot noise is still a serious problem for high-frequency sources.

Detailed calculations that are beyond the scope of these lectures show that a gravitational wave with frequency $2\Omega/2\pi$ in the range of 100 to 1,000 Hz

---

[24] In Advanced LIGO the destructive inteference in the absence of a gravitational wave is nearly but not quite complete, in order to make the response of the interferometer to the wave linear rather than quadratic in the wave's amplitude.

would have been detectable in the first run of LIGO if its amplitude $\overline{h}$ were greater than about 2×10^{-23}, while the present Advanced LIGO could detect a gravitational wave in this frequency range with amplitude 2.5 times smaller, with $\overline{h} \simeq 8 \times 10^{-24}$. It is expected that eventually LIGO will be sensitive to gravitational waves in this frequency range with $\overline{h}$ as small as 3.5×10^{-24}.

Thus our earlier rough estimate, that a coalescing neutron star binary at the distance of the Hulse–Taylor binary would give a strain of about 3×10^{-19}, means that such a coalescence could be detected by the present LIGO out to a distance of about 10^4 times the distance of the Hulse–Taylor binary pulsar, or roughly 60 Mpc. A more careful calculation gives 30 Mpc. The expected rate of neutron star coalescence in our galaxy is estimated to be in the range of about 5×10^{-4} to 10^{-6} per year.[25] Taking account of the estimated number of binary neutron stars per galaxy and the number of galaxies per volume centered on our galaxy, the same authors estimated a rate of coalescences per year and per volume in the range 4×10^{-9} to 2×10^{-5} per Mpc^3 per year. These estimates led to a widespread expectation that the advanced version of LIGO when complete would eventually be detecting the coalescence of neutron star binaries.

What happened was even more exciting. In 2015, in the first observing run of Advanced LIGO, two gravitational wave signals were observed, GW150914[26] and GW151226.[27] In both cases, coincidental chirps were detected at both of the LIGO interferometers, in Washington and Louisiana. They fit what would be expected from general relativity for the coalescence of a binary pair of black holes, followed by the subsequent "ringdown" of the merged black hole as it approaches equilibrium. From analysis of the time-development of the coalescence signals, the black holes were found to have initial masses $36^{+5}_{-4} M_\odot$ and $(29 \pm 4) M_\odot$ for GW150914, and $14.2^{+8.3}_{-3.7} M_\odot$ and $(7.5 \pm 2.3) M_\odot$ for GW151226. From comparison of the signals with numerical simulations of coalescence, the final black holes were found to have masses $(62 \pm 4) M_\odot$ and $20.8^{+6.1}_{-1.7} M_\odot$, respectively, with the missing mass radiated in gravitational waves. From the strength of the signals (a peak value of 1.0×10^{-21} for the relevant strain component $h_{11} - h_{22}$ for GW150914) it was calculated that the sources of GW150914 and GW151226 were at luminosity distances 410^{+160}_{-180} Mpc and 440^{+180}_{-190} Mpc, respectively. Not only was this the first direct detection of gravitational waves – the observation of GW150914 revealed the existence of a class of black holes considerably more massive than had been anticipated.

---

[25] V. Kalogera, R. Narayan, D. N. Spergel, and J. H. Taylor, *Astrophys. J.* **556**, 340 (2001).
[26] B. P. Abbott *et al.* (LIGO Scientific Collaboration and Virgo Collaboration), *Phys. Rev. Lett.* **116**, 061102 (2016).
[27] B. P. Abbott *et al.* (LIGO Scientific Collaboration and Virgo Collaboration), *Phys. Rev. Lett.* **116**, 241103 (2016).

At the time of writing in 2018, three more gravitational wave signals from coalescing black holes have been detected, described in the table[28] below:

Event	d_L (Mpc)	$-\Delta E$ ($M_\odot c^2$)	m_{chirp} ($M_\odot$)	m_1, m_2 ($M_\odot$)	m_{final} ($M_\odot$)
GW150914	440^{+160}_{-180}	3.0 ± 0.5	$28.2^{+1.8}_{-1.7}$	$35.4^{+5.0}_{-3.4}, 29.8^{+3.3}_{-4.3}$	$62.2^{+3.7}_{-3.4}$
GW151226	440^{+180}_{-190}	$1.0^{+0.1}_{-0.2}$	$8.9^{+0.3}_{-0.3}$	$14.2^{+8.3}_{-3.7}, 7.5^{+2.3}_{-2.3}$	$20.8^{+6.1}_{-1.7}$
GW170104	880^{+450}_{-390}	$2.0^{+0.6}_{-0.7}$	$21.1^{+2.4}_{-2.7}$	$31.2^{+8.4}_{-8.0}, 19.4^{+5.3}_{-5.9}$	$48.7^{+5.7}_{-4.6}$
GW170608	340^{+140}_{-140}	$0.85^{+0.07}_{-0.17}$	$7.9^{+0.2}_{-0.2}$	$12^{+7}_{-2}, 7^{+2}_{-2}$	$18.0^{+4.8}_{-0.9}$
GW170814	540^{+130}_{-219}	$2.7^{+0.4}_{-0.3}$	$24.1^{+1.4}_{-1.1}$	$30.5^{+5.7}_{-3.0}, 25.3^{+2.8}_{-4.2}$	$53.2^{+3.2}_{-2.5}$

As mentioned above, it had been anticipated that the first gravitational waves to be detected at Advanced LIGO would be from coalescing neutron stars, so it was a surprise that the first five gravitational wave signals were from binaries whose members were much too massive to be neutron stars, or anything but black holes. Where were the gravitational waves from coalescing neutron stars?

Finally, on August 17, 2017, such a signal, GW170817, was detected,[29] most clearly at the Hanford interferometer, and, after dealing with a noisy glitch, also at the Livingston interferometer, and much more weakly at the Virgo interferometer in Italy. The relative weakness of the Virgo signal provided an important clue to the source's location in the sky. As explained in the mathematical notes at the end of this section, each interferometer has four blind spots – it can detect no gravitational wave that arrives on a line of sight in the plane of the interferometer and midway between the interferometer's arms, or at an angle greater by one, two, or three right angles. Thus, although the weakness of the signal observed at Virgo limited the use of this signal in determining the intrinsic properties of the source, this weakness showed that the source was close to one of Virgo's blind spots.

Analysis of the LIGO signal showed that the gravitational waves came from the coalescence of orbiting bodies with individual masses $(1.36-1.60)M_\odot$ and $(1.17-1.36)M_\odot$, and total mass $2.74^{+0.04}_{-0.01}M_\odot$, consistent with neutron stars as

---

[28] This is a compressed version of a table available from Wikipedia, which gives references for all these signals. The date of each detection is indicated by the signal name; thus, GW170814 was detected on August 14, 2017. Here d_L is the luminosity distance, inferred from the observed signal strength; $-\Delta E$ is the total energy emitted as gravitational waves; m_{chirp} is the chirp mass defined by Eq. (2.3.13); m_1 and m_2 are the black hole masses before coalescence; and m_{final} is the mass of the black hole left after coalescence. Some of the results given in the table for the first two events were recalculated since the original publication.

[29] B. P. Abbott *et al.* (LIGO Scientific Collaboration, Virgo Collaboration, and other collaborations), *Astrophys. J.* **848**, L12 (2017). For details about the further analysis of the signal, see B. P. Abbott *et al.* (LIGO Scientific Collaboration and Virgo Collaboration), *Phys. Rev. X* **9**, 011001 (2019).

well as black holes or ordinary stars, but reaching separations too close for anything but neutron stars or black holes. From the intensity of the gravitational wave signal, the luminosity distance was calculated to be 40^{+8}_{-14} Mpc, much closer than for any of the earlier gravitational wave signals. Most excitingly, 1.7 seconds later a short-duration gamma ray burst, GRB170817A, was detected by the orbiting observatories Fermi and INTEGRAL. Such a signal was expected from the coalescence of neutron stars, but not of black holes. The gamma ray signal gave more precise information about the location in the sky of the source, and also showed that the speed of gravitational waves and that of electromagnetic waves are indistinguishable.

Less than a day later, a search with optical telescopes revealed the glowing remnant of this event,[30] in an outlying part of the galaxy NGC4993 at a distance 40 Mpc in the constellation Hydra. The composition of this debris was consistent with what would be expected from the coalescence of neutron stars. Evidence emerged suggesting the production of heavy elements in neutron star coalescence, including elements such as gold and platinum that caught the public's imagination. All these observations tended to confirm the picture of kilonovae described in the previous section, though argument continues. Observations about 110 days later with the Chandra X-ray telescope indicate that this kilonova has made a black hole.[31]

This event, and the earlier detection of gravitational waves from black hole coalescences, clearly mark the beginning of a new era in astronomy.[32]

* * * * *

The detailed analysis of these gravitational wave signals is too complicated to be thoroughly described here, in part because the strength of the gravitational fields in the merger of black holes and the more massive neutron stars requires the use of numerical methods, and also because at least for black holes the analysis must take into account their spins. What follows is a simplified analysis, using the quadrupole approximation described in Section 2.3, and ignoring black hole and neutron star spin. As we shall see, even for the gravitational wave signal from an orbiting binary in a circular orbit, the intensity of the received signal, on which we rely in estimating the distance of the binary, depends on four independent angles: the angles i and i' between the line of sight and the normals to the planes of the interferometer and the binary; the angle ϕ between arm 1 of the interferometer and the projection of the line of sight onto the plane of the interferometer; and a fourth angle α, which we take

---

[30] D. A. Coulter et al., Science **358**, 1556 (2017).
[31] D. Pooley, P. Kumar, J. C. Wheeler, and B. Grossan, Astrophys. Lett. **859**, L23 (2018).
[32] The limitations set by seismic noise on the observation of gravitational waves of low frequency may be overcome by the observation of the effect of gravitational waves on the arrival times of pulses from pulsars. As pointed out by S. Detweiler, Astrophys. J. **234**, 1100 (1979), this method is sensitive to gravitational waves with periods of the order of 1 to 10 years. Seismic noise can also be avoided by space-based interferometers, such as the LISA (Laser Interferometer Space Antenna) system proposed by the European Space Agency.

as the angle between the vectors perpendicular to the line of sight in the plane of the interferometer and in the plane of the binary. It is only through the analysis of general relativistic corrections to the wave form, which is beyond our scope here, that this can all be sorted out.

We can write the gravitational wave amplitude $h_{11} - h_{22}$ to which the interferometer is sensitive in terms of the amplitudes h_+ and $h_\times$ introduced earlier. Suppose we introduce a coordinate system (distinguished by a tilde) in which the gravitational wave is traveling in the 3-direction, and further choose coordinates so that the metric perturbation takes the form

$$\tilde{h}_{00} = \tilde{h}_{i0} = \tilde{h}_{0i} = \tilde{h}_{3i} = \tilde{h}_{i3} = 0$$
$$\tilde{h}_{11} = -\tilde{h}_{22} \equiv \tilde{h}_+, \quad \tilde{h}_{12} = \tilde{h} \equiv \tilde{h}_\times, \quad (2.4.17)$$

all these components depending only on $\tilde{x}_3$ and time. In the coordinate system we have been using, in which the interferometer is in the 1–2 plane, the gravitational wave will be traveling in some general direction

$$\hat{n}_1 = \sin i \cos \phi, \quad \hat{n}_2 = \sin i \sin \phi, \quad \hat{n}_3 = \cos i, \quad (2.4.18)$$

and will take the form

$$h_{ij} = R(i,\phi)_{ik} R(i,\phi)_{jl} \tilde{h}_{kl},$$
$$h_{00} = h_{0i} = h_{i0} = 0, \quad (2.4.19)$$

where $R(i,\phi)$ is the rotation

$$R(i,\phi) = \begin{pmatrix} \sin\phi & \cos\phi & 0 \\ -\cos\phi & \sin\phi & 0 \\ 0 & 0 & 1 \end{pmatrix} \begin{pmatrix} 1 & 0 & 0 \\ 0 & \cos i & \sin i \\ 0 & -\sin i & \cos i \end{pmatrix}, (2.4.20)$$

which takes the 3-axis into the direction $\hat{n}$. A straightforward calculation gives

$$h_{11} = (\sin^2\phi - \cos^2\phi \cos^2 i)\tilde{h}_+ + \sin 2\phi \cos i \, \tilde{h}_\times$$
$$h_{12} = h_{21} = -\sin 2\phi (1 + \cos^2 i) \tilde{h}_+/2 - \cos 2\phi \cos i \, \tilde{h}_\times$$
$$h_{13} = h_{31} = \sin 2i \cos\phi \, \tilde{h}_+/2 - \sin\phi \sin i \, \tilde{h}_\times$$
$$h_{22} = (\cos^2\phi - \sin^2\phi \cos^2 i)\tilde{h}_+ - \sin 2\phi \cos i \, \tilde{h}_\times$$
$$h_{23} = h_{32} = \sin\phi \sin 2i \, \tilde{h}_+/2 + \cos\phi \sin i \, \tilde{h}_\times$$
$$h_{33} = -\sin^2 i \, \tilde{h}_+. \quad (2.4.21)$$

(Both $\tilde{h}_+$ and $\tilde{h}_\times$ depend only on $\hat{n} \cdot \mathbf{x}$ and on time. The reader can easily check that $\partial_i h_{ij} = 0$ and $h_{ii} = 0$, so we are still in transverse-traceless gauge.) In particular, the amplitude in Eqs. (2.4.14) and (2.4.15) to which the interferometer is sensitive is

$$h_{11} - h_{22} = -\cos 2\phi \, (1 + \cos^2 i) \, \tilde{h}_+ + 2\sin 2\phi \, \cos i \, \tilde{h}_\times, \quad (2.4.22)$$

where again

$$\tilde{h}_+ \equiv \tilde{h}_{11} = -\tilde{h}_{22}, \quad \tilde{h}_\times \equiv \tilde{h}_{12} = \tilde{h}_{21}. \quad (2.4.23)$$

Equation (2.4.22) has the important immediate consequence that any interferometer has four blind spots, directions from which gravitational waves cannot be detected, characterized by lines of sight with $\cos i = 0$ and $\cos 2\phi = 0$. Looking back at Eq. (2.4.18), we see that in these cases the line of sight is in the plane of the interferometer and along a line midway between two arms – that is, ϕ equal to $\pi/4$ – or at an angle $\phi = 3\pi/4$, $5\pi/4$, or $7\pi/4$. As mentioned above, the *non-observation* of the gravitational wave signal GW170817 in one interferometer when a strong signal was observed in two other interferometers gave important information concerning the line of sight to this source, which was not provided by the strong signals at LIGO.

We record for future use that the 1-axis in the tilde coordinate system (which is used in the definition of $\tilde{h}_+$ and $\tilde{h}_\times$) has components in the interferometer-based coordinate system

$$\tilde{1}_j = R(i,\phi)_{j1} = (\sin\phi, -\cos\phi, 0), \qquad (2.4.24)$$

which is the unit vector in the plane of the interferometer orthogonal to the direction of travel of the gravitational wave.

In order to calculate the quantity (2.4.22) for a gravitational wave produced by a binary star, we now need to go beyond the calculation of gravitational wave power emission in the previous section. For simplicity, we consider a binary with zero eccentricity. In the inertial coordinate system of the binary (distinguished by an asterisk), in which the orbit is in the x^*–y^* plane, the separation vector has components

$$r_x^* = a\cos\Omega t^*, \qquad r_y^* = a\sin\Omega t^*, \qquad r_z^* = 0, \qquad (2.4.25)$$

where, according to Eq. (2.1.2), $\Omega = \sqrt{GM/a^3}$, and t^* is the time corresponding to time t on Earth. (For instance, if we ignore cosmological effects, $t^* = t - d$.) If there is an angle i' between the line of sight to the Earth and the normal to the plane of the orbit, then in the tilde coordinate system introduced above, in which the line of sight to the Earth is in the $\tilde{3}$-direction, the components of the separation vector are

$$\tilde{r}_1 = r_x^* \cos\alpha - r_y^* \cos i' \sin\alpha = a\cos\alpha\cos\Omega t^* - a\cos i' \sin\alpha \sin\Omega t^*,$$
$$\tilde{r}_2 = r_x^* \sin\alpha + r_y^* \cos i' \cos\alpha = a\sin\alpha\cos\Omega t^* + a\cos i' \cos\alpha \sin\Omega t^*,$$
$$\tilde{r}_3 = -r_y^* \sin i = -a\sin i' \sin\Omega t^*, \qquad (2.4.26)$$

where α is the angle between the vector in the plane of the interferometer orthogonal to the line of sight (used above to define the tilde coordinate system) and the vector in the plane of the orbit perpendicular to the line of sight. In the tilde coordinate system, the relevant components of the mass quadrupole moment $\tilde{Q}_{ij} \equiv \mu\tilde{r}_i\tilde{r}_j$ are

2.4 Gravitational Wave Detection: Coalescing Binaries

$$\tilde{Q}_{11}(t) - \tilde{Q}_{22}(t) = 2\mu a^2 \Big[-[1 + \cos^2 i'] \cos 2\alpha \cos 2\Omega t^*$$
$$+ 2 \sin 2\alpha \cos i' \sin 2\Omega t^* \Big],$$

$$\tilde{Q}_{12}(t) = 2\mu a^2 \Big[-\frac{1}{2}[1 + \cos^2 i'] \sin 2\alpha \cos 2\Omega t^* - \cos i' \cos 2\alpha \sin 2\Omega t^* \Big].$$
(2.4.27)

According to Eq. (2.3.A27), the components $\tilde{h}_+$ and $\tilde{h}_\times$ at a distance d from the source to be used in Eq. (2.4.22) are

$$\tilde{h}_\times(\mathbf{x}, t) = \frac{2G}{d} \ddot{\tilde{Q}}_{12}(t^*), \quad \tilde{h}_+(\mathbf{x}, t) = \frac{G}{d} \Big[\ddot{\tilde{Q}}_{11}(t^*) - \ddot{\tilde{Q}}_{22}(t^*) \Big]. \quad (2.4.28)$$

Using Eq. (2.4.28) in Eq. (2.4.27), and then using the result in Eq. (2.4.22), we have the strain component

$$h_{11} - h_{22} = A \cos(2\Omega t^*) + B \sin(2\Omega t^*), \quad (2.4.29)$$

where

$$A = -\frac{8\Omega^2 G \mu a^2}{d}(1 + \cos^2 i')\big[\cos\phi (1 + \cos^2 i) \cos 2\alpha - 2 \sin\phi \cos i \sin 2\alpha\big]$$
(2.4.30)

and

$$B = \frac{16\Omega^2 G \mu a^2}{d} \cos i' \big[\cos\phi (1 + \cos^2 i) \sin 2\alpha + 2 \sin\phi \cos i \cos 2\alpha\big].$$
(2.4.31)

As for any system of point particles whose motion is monochromatic with frequency Ω, the gravitational wave signal from a circularly orbiting binary with frequency Ω has frequency 2Ω.

Since there is no way to determine the arbitrary zero of time in Eq. (2.4.25), the comparison of Eq. (2.4.29) with observation only allows the determination of the modulus $A^2 + B^2$, as well as Ω. Using Eqs. (2.1.6) and (2.3.13), the measurement of Ω and $\dot\Omega$ allows us to calculate a^3/M and the chirp mass $\mu^{3/5} M^{2/5}$. This suffices to allow a calculation of the factor $\mu a^2 = [\mu M^{2/3}][a^3/M]^{2/3}$ in A and B, so with a guess at a plausible range of the angles i, ϕ, i', and α we can use the measured value of $A^2 + B^2$ to infer a plausible range of values for the distance d. This alone would not provide values for a or the individual masses, let alone the actual values of the angles i, ϕ, i', and α, but fortunately, as remarked in Section 2.1 for general binaries, relativistic corrections (here including effects of black hole spin) come to our aid, and allow a complete determination of all binary parameters.

Bibliography for Chapter 2

- D. Branch and J. C. Wheeler, *Supernova Explosions* (Springer-Verlag, Berlin, 2017).
- J. D. E. Creighton and W. G. Anderson, *Gravitational Wave Physics and Astronomy* (Wiley-VCH, Weinheim, 2011).
- Z. Kopal, *Close Binary Systems* (Chapman and Hall, London, 1959).
- Z. Kopal, *The Roche Problem* (Kluwer, Dordrecht, 1989).
- C. W. Misner, K. S. Thorne, and J. A. Wheeler, *Gravitation* (W. H. Freeman & Co., San Francisco, CA, 1972).
- S. Weinberg, *Gravitation and Cosmology* (John Wiley & Sons, New York, 1972).

3
The Interstellar Medium

The space between the stars in our galaxy is filled with matter. The density of interstellar matter is much lower than in a good laboratory vacuum, but there is a lot of space between the stars, and interstellar matter plays an important role in galaxies. It comes in various forms and temperatures. All but a few percent is un-ionized hydrogen and helium, with densities of order tens of atoms per cm^3 near the Sun. There are colder denser molecular clouds, pretty much in the galactic plane, with temperatures between about 50 K and 150 K, in which new stars form. There is a very hot ionized corona of highly ionized atoms, extending far outside the plane of the galaxy, with temperatures of several millions of degrees Kelvin and densities of order 10^{-2} to 10^{-4} atoms/cm^3. There are HII regions surrounding hot stars in the plane of the galaxy, chiefly of ionized hydrogen, with densities of order 10^3 to 10^4 atoms/cm^3. Much of the interstellar medium is pervaded with magnetic fields. Apart from hydrogen and helium, the interstellar matter contains "metals": mostly carbon, nitrogen, and oxygen, and also measurable quantities of lithium, sodium, magnesium, aluminum, silicon, phosphorus, sulfur, calcium, and iron. Their abundances are all somewhat less than what is believed to be the average cosmic abundances, especially for aluminum and calcium. There are grains of solid matter, which are believed to contain the missing metal atoms. These grains have dimensions of order 10^{-4} cm or less, and redden the light from more distant stars. To stir up the interstellar medium there are also starlight, supernovae shock waves, cosmic rays, and turbulence.

It is all quite complicated. In line with the spirit of these lectures, this chapter will not attempt a comprehensive survey of interstellar matter, but will instead take up a number of important topics that can be treated analytically. In Section 3.1 we consider the formation of emission and absorption spectral lines, which are ubiquitous in astronomy, and in particular continue to provide most of our information about interstellar matter. Section 3.2 describes HII regions, including Strömgren's lovely analytic treatment of their extent, and a simple calculation of their heating by a hot star. Section 3.3 deals with the important issue of cooling in the interstellar medium, and as an example discusses its application to the balance between heating and cooling in HII regions. In Section 3.4 we consider the formation of stars, using an updated version of a theory of Jeans, that although far from realistic nevertheless provides a language in which star formation continues to be discussed.

Where stars or black holes find themselves in clouds of interstellar matter, some of that matter will drizzle down onto the surface. The matter being accreted onto a star usually delivers significant angular momentum as well as mass to the star and therefore takes the form of a disk. In other cases, especially for black holes, when the matter of the cloud is pretty much at rest the accreting matter can take the form of a sphere. These two limiting cases of accretion are described in Sections 3.5 and 3.6. The treatment of accretion disks in Section 3.5 will be carried over to the next chapter, when we consider accretion disks around the massive black holes at the center of quasars.

Section 3.7 deals with the emission of low-frequency radiation, such as radio waves, from hot ionized interstellar gas, through the process of bremsstrahlung. The results presented here differ from those commonly cited.

3.1 Spectral Lines

Interstellar matter can produce emission lines when it is excited by neighboring stars, as in HII regions. It can also produce emission lines even from clouds in local thermal equilibrium in cases where the cloud is somewhat optically thick only near one or more discrete frequencies. And it can produce absorption lines in the spectra of sources on the far side of interstellar clouds. To deal with all of these cases, let us first recall Eq. (1.2.6) for time-independent radiation transport:

$$\hat{n} \cdot \nabla \ell(\nu, \hat{n}, \mathbf{x}) = -K(\nu, \mathbf{x})\ell(\nu, \hat{n}, \mathbf{x}) + J(\nu, \mathbf{x})/4\pi c. \qquad (3.1.1)$$

Here $\ell(\nu, \hat{n}, \mathbf{x}) \, d^2\hat{n} \, d\nu \, d^3x$ is the energy of the photons in a small volume d^3x around $\mathbf{x}$, having directions in a small solid angle $d^2\hat{n}$ around $\hat{n}$, and with frequencies in a small range from ν to $\nu + d\nu$. Also, $K(\nu, \mathbf{x})$ (previously called $\kappa_{\text{abs}}(\nu, \mathbf{x})\rho(\mathbf{x})$) is the net fraction of photon energy absorbed by the medium per distance traveled, and $J(\nu, \mathbf{x})$ (previously called $j(\nu, \mathbf{x})\rho(\mathbf{x})$) is the rate of radiant energy emission in all directions by the medium per volume and per frequency interval. We are ignoring photon scattering here, so we can limit ourselves to photons traveling in a fixed direction $\hat{n}$ toward the observer, and hence we can drop the argument $\hat{n}$, take $\mathbf{x} = s\hat{n}$, and replace the argument $\mathbf{x}$ with s. Then Eq. (3.1.1) becomes

$$\frac{d}{ds}\ell(\nu, s) = -K(\nu, s)\ell(\nu, s) + J(\nu, s)/4\pi c. \qquad (3.1.2)$$

In cases where the absorption represented by the coefficient $K(\nu, s)$ is negligible, the solution is trivial:

$$\ell(\nu, s) = \ell(\nu, s_1) + \frac{1}{4\pi c} \int_{s_1}^{s} J(\nu, s)\, ds, \tag{3.1.3}$$

where a subscript 1 denotes any convenient reference point along the line of sight, such as a point source or the far end of an interstellar cloud.

To treat cases where absorption is not negligible, we introduce the optical depth from the fixed reference point s_1 to a point s:

$$\tau(\nu, s) \equiv \int_{s_1}^{s} K(\nu, s')\, ds'. \tag{3.1.4}$$

It is elementary to check that Eq. (3.1.2) has the exact solution

$$\ell(\nu, s) = e^{-\tau(\nu,s)} \left[\ell(\nu, s_1) + \frac{1}{4\pi c} \int_{s_1}^{s} J(\nu, s') e^{\tau(\nu, s')}\, ds' \right]. \tag{3.1.5}$$

The integral over s' can be easily calculated if we assume that the ratio $J(\nu, s')/K(\nu, s')$ is independent of s', whether or not $J(\nu, s')$ and $K(\nu, s')$ are individually independent of s'. In this case, we can write Eq. (3.1.5) as

$$\ell(\nu, s) = e^{-\tau(\nu,s)} \left[\ell(\nu, s_1) + \left(\frac{J(\nu)}{4\pi c K(\nu)}\right) \int_{s_1}^{s} K(\nu, s') e^{\tau(\nu, s')}\, ds' \right]$$

$$= e^{-\tau(\nu,s)} \left[\ell(\nu, s_1) + \left(\frac{J(\nu)}{4\pi c K(\nu)}\right) \int_{s_1}^{s} \frac{d\tau(\nu, s')}{ds'} e^{\tau(\nu, s')}\, ds' \right]$$

and therefore

$$\ell(\nu, s) = e^{-\tau(\nu,s)} \ell(\nu, s_1) + \left(\frac{J(\nu)}{4\pi c K(\nu)}\right) \left[1 - e^{-\tau(\nu,s)}\right]. \tag{3.1.6}$$

We can use Eq. (3.1.6) more generally if we regard the quantity $J(\nu)/K(\nu)$ as a suitable average of $J(\nu, s')/K(\nu, s')$ over the ray path from s_1 to s.

Now we must see how to calculate $J(\nu)$ and $K(\nu)$. We consider the contribution of transitions $a \leftrightarrow b$ between two energy levels of some sort of atoms (or molecules) in the interstellar medium, with $E_a > E_b$. In 1917 Einstein[1] defined a quantity A_a^b as the rate at which an atom will spontaneously make a transition from an energy level E_a to a lower energy level E_b, emitting a photon of frequency near $\nu_{ab} = (E_a - E_b)/h$. The frequency of the emitted photon will not be precisely equal to ν_{ab}, both because the finite lifetime of the

---

[1] A. Einstein, *Phys. Z.* **18**, 121 (1917). For a textbook account, see S. Weinberg, *Lectures on Quantum Mechanics*, 2nd edn. (Cambridge University Press, Cambridge, 2015), Section 2.1. A difference in notation should be mentioned: In *Lectures on Quantum Mechanics* the indices n, m on the A and B coefficients refer to individual states, while here the indices a and b refer to energy levels, some of which may contain more than one individual state.

initial (and possibly also the final) energy level gives the frequency distribution a natural width, and also because thermal motion of the atoms produces a spread of Doppler-shifted frequencies. If the probability that the photon is emitted with frequency between ν and $\nu + d\nu$ is $\phi(\nu)\,d\nu$, sharply peaked at $\nu = \nu_{ab}$ with $\int \phi(\nu)\,d\nu = 1$, then the rate of photon emission per volume, per solid angle, and per frequency interval is $\phi(\nu)A_a^b n_a/4\pi$, where n_a is the number per volume of atoms in energy level E_a. It takes a photon a time ds/c to travel a distance ds, so the amount $J(\nu)/4\pi c$ of additional radiant energy emitted per distance traveled, per solid angle, and per frequency interval is $h\nu\,\phi(\nu)A_a^b n_a/4\pi c$, and hence

$$J(\nu) = h\nu\phi(\nu)A_a^b n_a. \tag{3.1.7}$$

Usually the natural width of atomic states is so small that the distribution of frequencies described by $\phi(\nu)$ arises chiefly from Doppler broadening. The observed frequency ν is related to the frequency ν_{ab} of the atomic transition at rest by

$$\nu = \nu_{ab}[1 - v/c], \tag{3.1.8}$$

where v is the atomic velocity along the line of sight. In thermal equilibrium at temperature T, where the mean velocity has a component $\bar{v}$ along the line of sight, the probability that a particle of mass m will have a velocity along the line of sight between v and $v + dv$ is $P(v)\,dv$, where

$$P(v) = \left(\frac{m}{2\pi k_B T}\right)^{1/2} \exp\left(-m(v - \bar{v})^2/2k_B T\right). \tag{3.1.9}$$

This is normalized so that $\int P(v)\,dv = 1$, so, to get a frequency distribution function $\phi(\nu)$ normalized so that $\int \phi(\nu)\,d\nu = 1$, we take

$$\phi(\nu) = \frac{c}{\nu_{ab}} P\big(c(1 - \nu/\nu_{ab})\big). \tag{3.1.10}$$

Einstein also considered the absorption of photons from radiation with an energy density $\rho_\gamma(\nu)\,d\nu$ at frequencies between ν and $\nu + d\nu$. The rate at which an individual atom in such a field makes a transition from an energy level E_b to a higher energy level E_a by absorbing a photon of frequency ν near $\nu_{ab} = (E_a - E_b)/h$ is written as $B_b^a \rho_\gamma(\nu)$. Again, the absorbed photon frequency is not generally exactly equal to ν_{ab}, because both the natural widths of the states and Doppler frequency shifts give the transition a finite width. If the probability that the photon is absorbed with frequency between ν and $\nu + d\nu$ is $\phi(\nu)d\nu$, then the rate of photon absorption per volume, per frequency interval, and per solid angle is $B_b^a \phi(\nu)\ell(\nu,s)n_b$ where n_b is the number density of atoms in energy level E_b, and as before $\ell(\nu, s)$ is the energy density at s per frequency interval, and per solid angle. Each photon absorption removes an energy $h\nu$ from the radiation field, so absorption reduces the energy density per solid angle and per frequency interval at a rate $B_b^a \phi(\nu)\ell(\nu,s)n_b$. Einstein also took into account the possibility that the radiation would stimulate the emission of photons by the atom in transitions from a higher energy level E_a to a lower level E_b at a rate written as $B_a^b \rho_\gamma(\nu)$, with the emitted photon frequency ν near ν_{ab}. The *net* rate of decrease of radiant energy density per solid angle and per frequency interval due to absorption and stimulated emission is then $(B_b^a n_b - B_a^b n_a)\phi(\nu)\ell(\nu, s)$. The net decrease $K(\nu)\ell(\nu, s)\,ds$ in radiant energy density per solid angle and per frequency interval due to absorption and stimulated emission during the time ds/c it takes a photon to travel a distance ds is $h\nu\phi(\nu)(B_b^a n_b - B_a^b n_a)\ell(\nu, s)\,ds/c$, so

$$K(\nu) = h\nu\phi(\nu)\left(B_b^a n_b - B_a^b n_a\right)/c. \qquad (3.1.11)$$

To work out the relations between the coefficients A_a^b, B_b^a, and B_a^b, Einstein imposed the condition that black-body radiation at temperature T should have no effect on the population of atomic states if the atoms are in thermal equilibrium at the same temperature. In thermal equilibrium the number densities of atoms of a given element in energy levels E_a and E_b have the ratio

$$n_a/n_b = \frac{g_a e^{-E_a/k_B T}}{g_b e^{-E_b/k_B T}}, \qquad (3.1.12)$$

where g_a and g_b are the degeneracies, the numbers of individual states having energies E_a and E_b, respectively. Einstein's condition (neglecting the widths of the states) is thus

$$\rho(\nu_{ab}, T)\left(B_b^a g_b e^{-E_b/k_B T} - B_a^b g_a e^{-E_a/k_B T}\right) = A_a^b g_a e^{-E_a/k_B T}, \qquad (3.1.13)$$

where $\rho(\nu, T)$ is the energy density per frequency interval for black-body radiation at temperature T

$$\rho(\nu, T) = \frac{8\pi h \nu^3}{c^3}\left[\exp(h\nu/k_B T) - 1\right]^{-1}. \qquad (3.1.14)$$

The coefficients A_a^b and B_a^b characterize atomic energy levels, not the atom's environment, and are therefore independent of temperature. In order for Eq. (3.1.14) to be satisfied for all temperatures with temperature-independent A and B coefficients, these coefficients must satisfy the Einstein relations:

$$B_b^a g_b = B_a^b g_a, \quad A_a^b = \left(\frac{8\pi h \nu_{ab}^3}{c^3}\right) B_a^b. \qquad (3.1.15)$$

Thus in general, whether or not the medium is in a state of thermal equilibrium, Eqs. (3.1.11) and (3.1.15) give

$$K(\nu) = h\nu\phi(\nu)\left[1 - \frac{n_a/g_a}{n_b/g_b}\right] B_b^a n_b/c \qquad (3.1.16)$$

and Eqs. (3.1.7), (3.1.15), and (3.1.16) give

$$\frac{J(\nu)}{K(\nu)} = \left(\frac{8\pi h \nu_{ab}^3}{c^2}\right) \left[\frac{n_b/g_b}{n_a/g_a} - 1\right]^{-1}. \qquad (3.1.17)$$

Now let us see how these general results work in special cases.

Emission Lines from Clouds in Thermal Equilibrium

In cases where an interstellar cloud is itself the source of an observed spectral line, we can take the reference point s_1 to be at the far edge of a cloud, where $\ell = 0$, so that Eq. (3.1.6) gives the observed radiation energy density per frequency interval and per solid angle at s as

$$\ell(\nu, s) = \left(\frac{J(\nu)}{4\pi c\, K(\nu)}\right) \left[1 - e^{-\tau(\nu,s)}\right]. \qquad (3.1.18)$$

If we assume that the cloud is in thermal equilibrium at temperature T, with number densities satisfying Eq. (3.1.12), then Eq. (3.1.17) gives

$$\frac{J(\nu)}{K(\nu)} = \left(\frac{8\pi h \nu_{ab}^3}{c^2}\right) \left[\exp(h\nu_{ab}/k_B T) - 1\right]^{-1}. \qquad (3.1.19)$$

Hence Eq. (3.1.18) gives

$$\ell(\nu, s) = \frac{\rho(\nu_{ab}, T)}{4\pi} \left[1 - e^{-\tau(\nu,s)}\right], \qquad (3.1.20)$$

where $\rho(\nu, T)$ is the energy density per frequency interval in black-body radiation at temperature T, given by Eq. (3.1.14). Also, in thermal equilibrium, Eqs. (3.1.4), (3.1.16), and (3.1.12) give the optical depth of the cloud due to transitions $a \leftrightarrow b$ as

$$\tau(\nu, s) = h\nu \phi(\nu) B_b^a \left[1 - \exp(-h\nu_{ab}/k_B T)\right] N_b(s)/c, \qquad (3.1.21)$$

where $N_b(s)$ is the column density of atoms in the lower energy level between s_1 and s

$$N_b(s) = \int_{s_1}^{s} n_b(s')\, ds'. \qquad (3.1.22)$$

If the cloud were optically thick, with $\tau(\nu, s) \gg 1$, at all frequencies, then the observer at s looking into the cloud would simply see black-body radiation at the cloud's temperature, with no spectral lines. (The denominator 4π in Eq. (3.1.20) is present because ℓ is the radiation energy density per frequency interval and per solid angle, while $\rho(\nu, T)$ is simply the black-body radiation energy density per frequency interval.) Commonly the optical thickness $\tau(\nu, s_1)$ is negligible except near one or more transition frequencies ν_{ab} where $\phi(\nu)$ is appreciable. The observer will then see a negligible radiant energy from the cloud except near these frequencies, which will appear as emission lines.

The most famous example of such an emission line is the 21 cm line of hydrogen, produced in transitions between the two hyperfine states of the $1s$ level of hydrogen, of which the lower has total (electron plus proton) spin $S = 0$ and the upper has total spin $S = 1$. These states are separated in energy by 5.9×10^{-6} eV, which is much less than the value of $k_B T$ even for the microwave background temperature 2.7 K, so the column densities of the states of the lower and upper state are proportional to the multiplicity $2S + 1$ of states for spin S, and thus these column densities are 1/4 and 3/4 of the total column density N_H of $1s$ atomic hydrogen, respectively. Here Eq. (3.1.21) gives an optical depth

$$\tau(\nu, s) = \frac{(h\nu_{ab})^2 \phi(\nu) B_b^a}{4 c k_B T} N_H(s). \tag{3.1.23}$$

The decay rate of the upper to the lower hyperfine state is $A_a^b = 2.87 \times 10^{-15}$ s^{-1}, from which it is possible to calculate the B coefficient using Eq. (3.1.15). The optical depth is then given by Eq. (3.1.23) as $\tau = 5.49 \times 10^{-14} N_H P(\nu)/T$, with the column density N_H in hydrogen atoms per cm^2, T in degrees Kelvin, $P(v)$ the velocity distribution function (3.1.9), and v related to ν by Eq. (3.1.8). The observed radiation energy density per solid angle and per frequency interval is given in terms of this optical depth by Eq. (3.1.20), which for small but non-negligible optical depth reads

$$\ell(\nu, s) = \frac{\rho(\nu_{ab}, T)}{4\pi} \tau(\nu, s). \tag{3.1.24}$$

The measurement in this way of the optical depth at the 21 cm wavelength provides information about the mean velocity, temperature, and column density of interstellar atomic hydrogen and eventually (it is hoped) of intergalactic atomic hydrogen.

Emission Lines from Non-equilibrium Regions

Of course interstellar matter that is not in thermal equilibrium can produce emission lines even (and especially) if it is optically thin. In this case, again taking s_1 to be the far end of the relevant interstellar matter, where $\ell = 0$, the radiation energy density is given by Eq. (3.1.3) as

$$\ell(\nu, s) = \left(\frac{J(\nu)}{4\pi c} \right) L(s), \tag{3.1.25}$$

where $L(s) \equiv s - s_1$ is the effective path length through the emitting region. As mentioned earlier, if the emission function J varies along this path, then its average should be used in Eq. (3.1.25).

Common examples of this sort of emission line include the optical-frequency lines emitted in the recombination of ionized interstellar matter, as in HII regions, discussed in the next section. Among these lines are those emitted by the decay of excited states of atomic hydrogen (HI), once-ionized and twice-ionized oxygen (OII and OIII), and once-ionized nitrogen (NiII), usually excited by ultraviolet radiation from a nearby hot star. A pair of green emission lines were once thought to indicate the presence of a new element, *nebulium*, until they were identified as arising from OIII. (The importance of OIII in cooling interstellar gas will be discussed in Section 3.3.) Much of our knowledge of Strömgren spheres comes from the observation of such emission lines.

It is not always the case that τ is very small. Where it is not, we must return to the general form of Eq. (3.1.6), which (when $\ell = 0$ vanishes at $s = s_1$) gives

$$\ell(\nu, s) = \left(\frac{J(\nu)}{4\pi c \, K(\nu)} \right) \left[1 - e^{-\tau(\nu,s)} \right], \qquad (3.1.26)$$

and use Eq. (3.1.16) for $K(\nu)$. In particular, where there is a population inversion, with $n_a/n_b > g_a/g_b$ (which never happens in thermal equilibrium), stimulated emission exceeds absorption, and we have K and τ *negative*, so that the interstellar material emits radiation at frequencies at which $-K(\nu)$ is large. Such masers are found at microwave frequencies in the accretion disks surrounding the black holes at the centers of various galaxies, including the well-studied cases of NGC 4258 and M33.

Absorption Lines

Absorption lines are produced in the passage of radiation from a distant source through interstellar matter. In this case, the relevant term in the observed radiation density is the first term in Eq. (3.1.6):

$$\ell(\nu, s) \simeq \ell(\nu, s_1) e^{-\tau(\nu, s)}, \qquad (3.1.27)$$

where s_1 is the coordinate of the source. Absorption lines occur at values $\nu_{ab} = (E_a - E_b)/h$ of frequency where the optical depth τ has a peak, due to transitions between the energy levels a and b. When the medium is in thermal equilibrium, the optical depth is given by Eq. (3.1.21).

The absorption line at 21 cm wavelength is seen in the spectrum of radio sources viewed through clouds of interstellar hydrogen, and used to measure the velocity, temperature, and column density of such clouds. As we shall see in the following chapter, these observations were used to map the rotational speeds at various positions in distant disk galaxies, and in that way provided one of the first pieces of evidence for dark matter in galaxies. It should be noted that at radio wavelengths the optical depth is reduced by stimulated emission in the transition $a \to b$ from the upper to the lower state. In particular, for $k_B T \gg h\nu_{ab}$, Eq. (3.1.21) shows that the optical depth is reduced by stimulated emission by a factor $h\nu_{ab}/k_B T$:

$$\tau(\nu, s_1) = h\nu \phi(\nu) B_b^a \left(h\nu_{ab}/k_B T \right) N_b / c. \qquad (3.1.28)$$

Absorption lines at optical wavelengths are seen in the spectra of stars viewed through interstellar clouds. There is a classic instance of interstellar absorption at optical wavelengths in the spectrum of ζ Ophiuchi. On the line of sight between this star and the Earth there are clouds of cold gas that absorb light at various optical frequencies of atoms and molecules. For instance, there is absorption at the 5,890 Å and 5,896 Å lines of atomic sodium, produced in transitions of the $3s_{1/2}$ state to the $3p_{3/2}$ and $3p_{1/2}$ states. (These are the transitions that in 1925–1926 gave the first evidence of electron spin.) From the Doppler shift of these lines, it is found that two of the thicker clouds have velocities along the line of sight of 30 km/sec and 10 to 20 km/sec, respectively.[2]

Though it was not realized at the time, an optical absorption line in the spectrum of ζ Oph provided the first evidence for the existence of the cosmic microwave background.[3] There had been a measurement of the radiation temperature at wavelength 0.264 cm in 1941, long before the discovery by Penzias and Wilson. In 1941 W. S. Adams,[4] following a suggestion of Andrew McKellar, found two dark lines in the spectrum of ζ Oph that could be identified as due to absorption of light by cyanogen (CN) in an intervening cloud. The first line, observed at a wavelength of 3,874.62 Å, could be attributed to absorption of light from the CN ground state, with rotational angular momentum $J=0$, leading to the component of the first vibrationally excited state with $J=1$. But the second line, at 3,874.00 Å, represented absorption from the $J=1$ rotationally excited vibrational ground state, leading to the $J=2$ component of the first vibrationally excited state.[5] From this, McKellar concluded[6] that a fraction of the CN molecules in the cloud were in the first excited rotational component of the vibrational ground state, which is above the true $J=0$ ground state by an energy $hc/(0.264 \text{ cm})$, and from this fraction he estimated an equivalent molecular temperature of 2.3 K. Of course, he did not know that the CN molecules were being excited by radiation, much less by black-body radiation. After the discovery by Penzias and Wilson several astrophysicists[7] independently noted that the old Adams–McKellar result could be explained by radiation with a black-body temperature at wavelength 0.264 cm in the neighborhood of 3 K. Theoretical analysis showed that nothing else could explain the

---

[2] M. J. Barlow et al., *Mon. Not. Roy. Astron. Soc.* **272**, 333 (1995).
[3] The following discussion is based on that of S. Weinberg, *Cosmology* (Oxford University Press, New York, 2008), Section 2.1.
[4] W. S. Adams, *Astrophys. J.* **93**, 11 (1941).
[5] Today the wavelengths of these two lines are more accurately known to be 3,874.608 and 3,873.998 Å. There is another line at 3,875.763 Å, produced by transitions from the $J=1$ rotationally excited vibrational ground state to the $J=0$ component of the first vibrationally excited state.
[6] A. McKellar, *Publs. Dominion Astrophys. Observatory (Victoria, B.C.)* **7**, 251 (1941).
[7] G. Field, G. H. Herbig, and J. L. Hitchcock, *Astron. J.* **71**, 161 (1966); G. Field and J. L. Hitchcock, *Phys. Rev. Lett.* **16**. 817 (1966); *Astrophys. J.* **146**, 1 (1966); N. J. Woolf, quoted by P. Thaddeus and J. F. Clauser, *Phys. Rev. Lett.* **16**, 819 (1966); I. S. Shklovsky, *Astronomicheskii Tsirkular* No. **364** (1966).

excitation of this rotational state.[8] This interpretation was then borne out by continuing observations on this and other absorption lines in CN as well as CH and CH$^+$ in the spectrum of ζ Oph and other stars.

3.2 HII Regions

Here and there throughout the plane of the galaxy there are hot stars at the upper end of the main sequence, of spectral type O and B, that emit copious photons with energies so high that they can ionize the atomic hydrogen (HI) that surrounds these stars. These photons cannot readily be absorbed by ionized hydrogen,[9] so they travel to the outskirts of the ionized region, where they ionize more hydrogen. At some distance from the star the radiation becomes attenuated enough that its production of hydrogen ions (HII) no longer overwhelms the recombination of protons and electrons, and the fraction of un-ionized hydrogen begins to increase. As soon as there is any appreciable amount of HI, the interstellar medium becomes opaque to ionizing radiation, and at even slightly greater distances there is no more ionization. In this way the hot star's ultraviolet radiation excavates a sphere of ionized hydrogen around the star, with a remarkably distinct outer surface. The sphere is observed through its production of radiation emitted when protons and electrons recombine into excited states of atomic hydrogen, which then decay to the ground state through a cascade of intermediate states. This notably includes the Lyman-α ultraviolet line emitted when hydrogen atoms make the final transition from the $2p$ state to the $1s$ ground state. Such spheres were first discussed by the Danish astronomer Bengt Strömgren[10] (1908–1987), and are widely known as Strömgren spheres.

To analyze the structure of Strömgren spheres, we will consider only the simplest case: a spherically symmetric star in a medium that in the absence of the star would be a uniform gas of pure atomic hydrogen, of constant number density n. If at a distance r from the star's center a fraction $\xi(r)$ of the hydrogen is not ionized by the star's ultraviolet radiation, then the number density of hydrogen atoms is $\xi(r)n$, and the number densities of protons and electrons are both equal to $(1 - \xi(r))n$. If the star emits $\mathcal{L}$ photons per time isotropically in all directions, then at r the flux $\Phi(r)$ (in photons per area per time) is given by

$$\Phi(r) = \frac{\mathcal{L}}{4\pi r^2} \exp\bigl(-\tau(r)\bigr). \qquad (3.2.1)$$

---

[8] G. Field, G. H. Herbig, and J. L. Hitchcock, *Astron. J.* **71**, 161 (1966); Thaddeus and J. F. Clauser, *Phys. Rev. Lett.* **16**, 819 (1966).

[9] Free–free absorption of photons in the Coulomb scattering of electrons and protons is less important in HII regions, and will not be taken into account in this section.

[10] B. Strömgren, *Astrophys. J.* **89**, 526 (1939).

Here $\tau(r)$ is the optical depth between r and the star's surface at the stellar radius r_0:

$$\tau(r) = \int_{r_0}^{r} n\sigma\xi(r')\,dr', \qquad (3.2.2)$$

where σ is the photoionization cross section in atomic hydrogen, with its frequency dependence averaged over the star's spectrum. The number of ionizations per volume per time at r is $\xi(r)n\sigma\Phi(r)$, while the number of recombinations per volume per time is the product $n^2(1-\xi(r))^2$ of the number densities of protons and electrons times a coefficient α with the dimensions of length^3 per time, so the condition for equilibrium which determines the fractional ionization at r is

$$\xi(r)n\sigma\Phi(r) = \alpha n^2(1-\xi(r))^2. \qquad (3.2.3)$$

(We are treating α as a constant. In principle α depends on the temperature of the plasma of electrons and protons, but the temperature does not vary sharply within the Strömgren sphere, and since recombination is an exothermic process its temperature dependence is not pronounced.[11])

From these relations we can derive a differential equation for $\xi(r)$. From Eqs. (3.2.1) and (3.2.3) we have

$$\exp(\tau(r)) = \frac{\mathcal{L}\sigma}{4\pi r^2 \alpha n} \frac{\xi(r)}{(1-\xi(r))^2}. \qquad (3.2.4)$$

Differentiating with respect to r, using Eq. (3.2.2) for $d\tau/dr$, and then dividing by e^τ gives

$$r^2 \frac{d}{dr}\left(\frac{\xi(r)}{r^2(1-\xi(r))^2}\right) = n\sigma \frac{\xi^2(r)}{(1-\xi(r))^2}. \qquad (3.2.5)$$

The only astrophysical parameter on which this differential equation depends is $n\sigma$, but the other parameters in the problem get into the solution through the initial condition. Since $\tau(r_0)$ vanishes by definition, at r_0 Eq. (3.2.4) gives

$$\frac{\xi(r_0)}{(1-\xi(r_0))^2} = \frac{4\pi r_0^2 \alpha n}{\mathcal{L}\sigma}. \qquad (3.2.6)$$

Equation (3.2.5) is too complicated to allow a simple analytic solution. We can, however, find such solutions in two overlapping regions.

---

[11] For temperatures in a range of 5,000 K to 20,000 K, the recombination coefficient α varies from 4.54×10^{-13} cm^3/sec to 2.52×10^{-13} cm^3/sec. This is for "case B" recombination, in which the ultraviolet photons emitted in recombination directly to the ground state of hydrogen are supposed immediately to ionize a nearby hydrogen atom, so that recombination to the ground state is not included in the effective value of α. Numerical estimates in this section are taken from D. E. Osterbrock, *Astrophysics of Gaseous Nebulae and Active Galactic Nuclei* (University Science Books, Mill Valley, CA, 1989).

Interior of the Sphere

By definition the interior of a Strömgren sphere is a region in which the ionization is nearly 100%, so that ξ is much less than unity. Where this is the case we can set $1 - \xi$ equal to unity in Eq. (3.2.5), which then becomes

$$r^2 \frac{d}{dr} \frac{\xi(r)}{r^2} = n\sigma \xi^2(r). \tag{3.2.7}$$

The solution is now elementary:

$$\frac{\xi(r)}{r^2} = \left[r_0^2 / \xi(r_0) - n\sigma(r^3 - r_0^3)/3 \right]^{-1}. \tag{3.2.8}$$

This is valid for values of r for which it gives $\xi(r) \ll 1$.

To see when this condition is satisfied, we note that two independent and very different distance scales enter in Eq. (3.2.8). One is the mean free path d of photons in a gas of atomic hydrogen of number density n:

$$d \equiv 1/\sigma n. \tag{3.2.9}$$

The other distance scale is the radius R_S that a sphere of completely ionized hydrogen of density n would need in order for there to be one recombination for every photon emitted by the star, if placed at the center of the sphere:

$$\alpha n^2 \left(\frac{4\pi R_S^3}{3} \right) \equiv \mathcal{L}. \tag{3.2.10}$$

For typical Strömgren spheres R is in the range of 10–100 pc, while d is much smaller, typically of order 0.1 pc.

For $\xi \ll 1$, Eqs. (3.2.6), (3.2.9), and (3.2.10) give $\xi(r_0)/r_0^2 = 3d/R^3$, so Eq. (3.2.8) may be rewritten as

$$\xi(r) = \frac{3dr^2}{R_S^3 - r^3 + r_0^3}.$$

The stellar radius r_0 is negligible compared with R or even d, so we do not need to keep the term r_0^3 in the denominator of this formula, which becomes

$$\xi(r) = \frac{3dr^2}{R_S^3 - r^3}. \tag{3.2.11}$$

We see that as long as r is no larger than of order R_S, but not close to R_S, $\xi(r)$ is no larger than of order $d/R_S \ll 1$. Under these circumstances Eqs. (3.2.8) and (3.2.11) should be good approximations, and we can conclude that the Strömgren sphere has a radius close to R_S. For instance, according to Eq. (3.2.11) ξ rises to 0.1 when $r \simeq R_S - 10d$, which for typical Strömgren spheres is very close to R_S.

Surface of the Sphere

Equation (3.2.11) shows that the approximation $\xi \ll 1$ must break down when r is within some distance of order d of the nominal radius R_S. For r in this range we can no longer set the factor $1 - \xi$ equal to unity in Eq. (3.2.5), but we can get an analytic solution with a different approximation. Since this range is so narrow compared with R_S, we can take r in Eq. (3.2.5) as the constant R_S, so that Eq. (3.2.5) becomes

$$\frac{d}{dr}\left(\frac{\xi(r)}{(1-\xi(r))^2}\right) = n\sigma \frac{\xi^2(r)}{(1-\xi(r))^2}. \tag{3.2.12}$$

Then if $\xi(r)$ grows from a small value ξ_a at a radius r_a to a value ξ_b close to unity at radius r_b, we have

$$\frac{r_b - r_a}{d} = \int_{\xi_a}^{\xi_b} \left[\frac{(1-\xi)^2}{\xi^2}\right] \frac{d}{d\xi}\left(\frac{\xi}{(1-\xi)^2}\right) d\xi$$

$$= \left[-\frac{1}{\xi} + 2\ln\left(\frac{\xi}{1-\xi}\right)\right]_{\xi_a}^{\xi_b}. \tag{3.2.13}$$

For instance, ξ will rise from 0.1 to 0.99 in a range of length $22.6d$ near R_S. For typical Strömgren spheres this is much less than the radius R_S of the sphere, showing that the transition from ionized to atomic hydrogen does indeed take place in a very narrow shell.

Recombination Lines

According to the definition of the recombination coefficient α, the energy of the photons emitted per second and per volume by recombination with frequency between ν and $\nu + d\nu$ is

$$J(\nu, s)\, d\nu = \alpha n^2(s)\big(1 - \xi(s)\big)^2 h\nu \phi(\nu)\, d\nu, \tag{3.2.14}$$

where s is the coordinate along the line of sight, and $\phi(\nu)$ is the line shape function introduced in the previous section, with $\int \phi(\nu)\, d\nu \equiv 1$. Neglecting absorption, assuming total ionization within the HII region, and assuming no radiation at frequency ν coming into the HII region at its far end, Eq. (3.1.3) gives the intensity of the line at frequency ν leaving the region toward the observer as

$$\ell(\nu) = \frac{\alpha \mathcal{E} h\nu \phi(\nu)}{4\pi c}, \tag{3.2.15}$$

where $\mathcal{E}$ is the so-called *emission measure*

$$\mathcal{E} = \int_{\text{HII region}} n^2(s)\, ds. \tag{3.2.16}$$

The observation[12] of the Hα ($n = 3 \to n = 2$) and Hβ ($n = 4 \to n = 2$) hydrogen emission lines indicates that typical HII regions have $\mathcal{E} \approx 10^3 – 10^4$ pc cm^{-6}, so such an HII region with a diameter of, say, 10 pc would have a number density of order 10–30 cm^{-3}.

Heating

HII regions provide a particularly simple example of the heating of interstellar matter by starlight. First note that the collision of photons with electrons is a very ineffective way of transferring energy from radiation to matter. As noted in Section 1.4, energy and momentum conservation do not allow a photon to be absorbed by a free electron in empty space, while in the scattering of a photon of energy E at an angle θ by an electron at rest, the energy transferred to the electron is

$$\Delta E = \frac{E^2(1+\cos^2\theta)}{m_e c^2 - E(1+\cos^2\theta)},$$

which is very much less than E for optical or ultraviolet frequency photons, for which $E \ll m_e c^2$. Scattering by a free proton is even less effective. As long as there is any appreciable amount of neutral hydrogen present, the dominant mechanism by which a hot O or B star can heat the interstellar medium is photoionization.

HII regions are typically sufficiently wide that any photon emitted by the central star with an energy E above the ionization threshold $E_I = 13.6$ eV of neutral hydrogen will ionize some hydrogen atom, giving the emitted electron an energy $E - E_I$, so the average energy per electron given to these electrons will be

$$\overline{\Delta E} = \frac{\int_{E_I}^\infty (E - E_I)\mathcal{L}(E)\,dE}{\int_{E_I}^\infty \mathcal{L}(E)\,dE}, \qquad (3.2.17)$$

where $\mathcal{L}(E)\,dE$ is the rate at which the star emits photons with energy between E and $E + dE$. To a good approximation, the energy dependence of $\mathcal{L}(E)$ is that of a black body with some "color temperature" T_c:

$$\mathcal{L}(E) \propto \frac{E^2}{\exp(E/k_B T_c) - 1}. \qquad (3.2.18)$$

It is easy to evaluate $\overline{\Delta E}$ in two limiting cases. For $k_B T_c \ll E_I$, the integrals in Eq. (3.2.17) are dominated by energies just above the ionization threshold E_I, so, by setting $E = E_I + w$, we find

$$\overline{\Delta E} \simeq \frac{\int_0^\infty w E_I^2 \exp(-(E_I + w)/k_B T_c)\,dw}{\int_0^\infty E_I^2 \exp(-(E_I + w)/k_B T_c)\,dw} = k_B T_c. \qquad (3.2.19)$$

---

[12] Cited by W. J. Maciel, *Astrophysics of the Interstellar Medium* (Springer, New York, 2013).

For $k_B T \gg E_I$, the integrals in Eq. (3.2.17) are dominated by energies far above the ionization threshold E_I, so we can ignore the lower bound in these integrals, and find

$$\overline{\Delta E} \simeq \frac{\int_0^\infty [E^3/(\exp(E/k_B T_c) - 1)]\, dE}{\int_0^\infty [E^2/(\exp(E/k_B T_c) - 1)]\, dE}$$

$$= \frac{\pi^4}{30 \zeta(3)} k_B T_c = 2.701 k_B T_c. \qquad (3.2.20)$$

Hence the order of magnitude of $\overline{\Delta E}$ is never very different from $k_B T_c$. Numerical calculations[13] show that $\overline{\Delta E}$ rises from $1.051\, k_B T_c$ for $k_B T_c/E_I = 0.033$ to $1.655 k_B T_c$ for $k_B T_c/E_I = 1.85$.

In equilibrium the constancy of the degree of ionization requires that the rate of ionizations per volume must equal the rate $\alpha n_e n_p$ of recombinations per volume. Hence the *heating function* Γ, the rate per volume at which photons deposit energy in the matter of the HII region, is

$$\Gamma = \alpha n_e n_p \overline{\Delta E}. \qquad (3.2.21)$$

Thermal equilibrium requires that the temperature of the medium must take a value at which this heating is balanced by some sort of cooling. There is one sort of cooling that clearly will not do the trick. If the average energy of electrons that recombine with protons is $\overline{E}_R$, then the rate per volume at which the ionized gas loses energy through recombination is $\alpha n_e n_p \overline{E}_R$. For this to balance the heating function (3.2.21), it would be necessary to have $\overline{E}_R = \overline{\Delta E}$. The cross section for electrons to be captured by protons decreases with increasing electron energy, so $\overline{E}_R$ is *less* than the mean electron energy $3k_B T_e/2$, where T_e is the electron kinetic temperature. We conclude that if recombination were the sole form of cooling, the electron temperature T_e would have to be greater than $2\,\overline{\Delta E}/3k_B$, which is always greater than $2T_c/3$. But HII regions excited by stars with color temperatures of 3×10^4 K to 5×10^4 K generally have electron temperatures less than around 10^4 K, considerably less than $2T_c/3$. Other cooling processes aside from recombination are needed to bring the temperature down this low. Such general cooling processes are considered in the next section.

3.3 Cooling

The interstellar medium is heated by starlight and supernova shock waves, so it can reach equilibrium only at a temperature at which this heating is balanced over time by processes of cooling. Furthermore, as we will see in the next

---

[13] L. Spitzer, Jr., *Physical Processes in the Interstellar Medium* (John Wiley & Sons, New York, 1998), Table 6.1.

section, in order for stars to form in the gravitational collapse of clumps of interstellar matter, it is necessary for gas pressure in these clumps to be reduced by cooling. So it is important to consider how to calculate the quantity known as the *cooling function* Λ, the rate per volume and per time at which interstellar matter loses energy. It is a function of the temperature and of the densities of whatever constituents of the interstellar medium participate in the cooling process.

Cooling typically occurs when kinetic energy in a collision is transferred to excited states of atoms, ions, or molecules, after which the excited states lose their energy by emitting photons, which escape the matter if it is optically thin. We will separately consider two limiting cases of such cooling processes, for which the cooling function has very different dependences on the density of the medium being cooled.

We will also consider a third source of cooling, which becomes dominant in the absence of metals, at temperatures high enough that almost all hydrogen and helium atoms are ionized. This is bremsstrahlung, the radiation of photons in the Coulomb scattering of electrons by ions. Bremsstrahlung is the inverse of the process leading to free–free absorption of photons, calculated in the appendix to Section 1.4, which is often called inverse-bremsstrahlung. In Section 3.7, bremsstrahlung will be discussed again as a source of observable low-frequency radiation.

Prompt Radiation

In some cases radiative decays are so rapid that a particle (an ion, atom, or molecule) that is excited in a collision will almost always lose its energy in radiation, rather than in a subsequent collision. Assuming that the radiation escapes the medium, the rate of energy loss per cm^3 per second is

$$\Lambda = \sum_{ab} n_a \Gamma_{ab} (E_b - E_a), \qquad (3.3.1)$$

where n_a is the number density of these particles (ions, atoms, or molecules) normally in energy level a with energy E_a, and Γ_{ab} is the rate at which collisions cause one of these particles to make a transition from energy level a to a higher energy level b. The transition rate Γ_{ab} is proportional to the density of whatever other projectile particles are colliding with the target particle in question, so in the case of prompt radiation the cooling function is quadratic in particle densities. As long as we know that the rate for the de-excitation transition $b \to a$ is dominated by radiation rather than collisions, to calculate the cooling function we do not need to know the radiative transition rate, but we do need to know the excitation rate Γ_{ab}.

In regions of interstellar space with low number density, collisions are mostly long-range affairs, so the largest contributions to collision rates are generally those for which the interaction energy falls off least steeply with distance. The electrostatic interaction energy between a projectile particle and target particle that produces a transition $a \to b$ in the target particle has a dependence on the separation R of the target and projectile of the form $R^{-\ell-\ell'-1}$, where ℓ and ℓ' are the dimensionalities in powers of length of the operators acting on the states of the target particle and projectile particle, respectively.[14] For monopoles, dipoles, quadrupoles, etc. we have $\ell = 0$, $\ell = 1$, $\ell = 2$, etc. The monopole operator is the total electric charge. Since we are interested in a transition $a \to b$ between different energy levels of the target particle, which cannot be produced by the total charge operator, the operator acting on the target must have $\ell \geq 1$, whether the target particle is charged or neutral. On the other hand, if we are not interested in transitions between energy levels of the projectile particle, the fall off of the interaction with distance is least steep if the projectile is a charged particle, such as an electron or an ion, in which case we can have $\ell' = 0$. At any given temperature electrons move much faster than ions, and hence have larger collision rates, so as long as electrons are plentiful it is their collisions with atoms, ions, etc. that play the leading role in cooling. The Coulomb attraction between electrons and positive ions increases the density of electrons near the ions, which adds to the effectiveness of electron–ion collisions. The recombination of ions and electrons after photoionization is generally a slow process, so ions and electrons can be plentiful in the interstellar medium even at temperatures far below the level at which they would cease to be abundant in thermal equilibrium.

Electron–electron scattering is generally sufficiently rapid that electrons are in kinetic equilibrium at some temperature T_e, with the number density of electrons with kinetic energy between E and $E + dE$ given by the Maxwell–Boltzmann distribution

$$n_e(E)\, dE = \frac{2 n_e \sqrt{E}\, dE}{\sqrt{\pi}(k_B T_e)^{3/2}} \exp\left(-E/k_B T_e\right),$$

where n_e is the total electron number density. If the rate at which an electron of energy $E \geq E_b - E_a$ produces a transition $a \to b$ between energy levels of target particles is $n_a \gamma_{ab}(E)$, then

$$\Gamma_{ab} = \int_{E_b - E_a}^{\infty} n_e(E)\, \gamma_{ab}(E)\, dE,$$

and so the cooling function (3.3.1) is

$$\Lambda = \frac{2 n_e}{\sqrt{\pi}(k_B T_e)^{3/2}} \sum_{ab} n_a (E_b - E_a) \int_{E_b - E_a}^{\infty} \gamma_{ab}(E) \sqrt{E} \exp\left(-E/k_B T_e\right) dE.$$
(3.3.2)

---

[14] For a textbook discussion, see S. Weinberg, *Lectures on Quantum Mechanics*, 2nd edn. (Cambridge University Press, Cambridge, 2015), Section 5.9.

In particular, if there is just one available transition $a \to b$, and the temperature is much less than $(E_b - E_a)/k_B$, then the integral in Eq. (3.3.2) will be dominated by values of E close to the lower bound $E_b - E_a$. The coefficient $\gamma_{ab}(E)$ will vanish in this limit like some power ν of $E - E_b + E_a$. That is, for $E \to E_b - E_a$

$$\gamma_{ab}(E) \to C_{ab}(E - E_b + E_a)^\nu,$$

with C_{ab} some constant. The cooling function (3.3.2) is then

$$\Lambda = \frac{2n_e \Gamma(\nu + 1)}{\sqrt{\pi}} n_a (k_B T_e)^{\nu - 1/2} (E_b - E_a)^{3/2} C_{ab} \exp\left(-(E_b - E_a)/k_B T_e\right). \tag{3.3.3}$$

For instance, neutral hydrogen atoms are by far the most abundant constituent of the interstellar medium, but the first excited level in atomic hydrogen is the $n = 2$ level, 10.2 eV above the $n = 1$ ground state. In this case, for T between about 4,000 K and 12,000 K, Eq. (3.3.3) gives[15]

$$\Lambda \simeq \Lambda_0 n_e n_{HI} \exp(-118{,}000\,\text{K}/T), \tag{3.3.4}$$

where $\Lambda_0 = 7.3 \times 10^{-19}$ erg cm^3/sec. This gives a rapid increase with temperature up to about 10,000 K, but this increase does not continue with higher temperature, because the density n_{HI} of un-ionized hydrogen begins to fall off rapidly for temperatures above about 15,000 K.

To understand this, note that in the interstellar medium far from sources of ionizing radiation the degree of ionization is set by a balance between collisional ionization and recombination.[16] As already mentioned in Section 3.2, the rate per time and per volume of recombination of electrons and HII ions (i.e., protons) into neutral hydrogen atoms takes the form

$$\Gamma(e + \text{HII} \to \text{HI} + \gamma) = \alpha(T) n_e n_{\text{HII}}, \tag{3.3.5}$$

where n_e and n_{HII} are the number densities of electrons and hydrogen ions, and $\alpha(T)$ is a coefficient with only power-law dependence on temperature. Detailed calculations give[17]

$$\alpha(T) \simeq 4.0 \times 10^{-13} \left(T\,[10^4\,\text{K}]\right)^{-0.6353} \text{cm}^3/\text{sec}, \tag{3.3.6}$$

---

[15] See e.g. L. Spitzer, Jr., *Physical Processes in the Interstellar Medium* (John Wiley & Sons, New York, 1998), Eq. (6-12).

[16] For studies of departures from this equilibrium, see O. Gnat and A. Sternberg, *Astrophys. J. Suppl.* **168**, 213 (2007), and earlier work cited therein.

[17] H. Mo, F. van den Bosch, and S. White, *Galaxy Formation and Evolution* (Cambridge University Press, Cambridge, 2010), Eq. (B1.50). This is for "Case A" recombination, in which the photon emitted in recombination is not supposed to ionize another neutral hydrogen atom, which is the case for instance in optically thin regions or if this photon is absorbed by a dust grain.

for T between about $100\,\text{K}$ and $10^8\,\text{K}$. On the other hand, the rate per time and per volume of collisional ionization takes the form (in a non-conventional notation)

$$\Gamma(e + \text{HI} \to \text{HII} + e + e) = \beta(T) n_e n_{\text{HI}} \exp(-T_\text{I}/T), \quad (3.3.7)$$

where $T_\text{I} \equiv 13.6\,\text{eV}/k_\text{B} = 157{,}800\,\text{K}$, and $\beta(T)$ is another coefficient with only power-law dependence on temperature. Detailed calculations give[18]

$$\beta(T) \simeq 2.32 \times 10^{-8} (T/T_\text{I})^{1/2} \left(1 + (T\,[10^5\,\text{K}])^{1/2}]\right)^{-1}\,\text{cm}^3\,\text{sec}^{-1}, \quad (3.3.8)$$

for T between about $10^4\,\text{K}$ and $10^8\,\text{K}$. In the common case of collisional ionization equilibrium, the rates per time and per volume of appearance and disappearance of neutral hydrogen atoms must be equal, so

$$\Gamma(e + \text{HII} \to \text{HI} + \gamma) = \Gamma(e + \text{HI} \to \text{HII} + e + e).$$

This gives a ratio of ionized and un-ionized hydrogen densities

$$\frac{n_{\text{HII}}}{n_{\text{HI}}} = \left(\frac{\beta(I)}{\alpha(T)}\right) \exp(-T_\text{I}/T),$$

or in other words

$$n_{\text{HI}} = n_\text{H}\left(1 + \exp(-T_\text{I}/T)\beta(T)/\alpha(T)\right)^{-1}, \quad (3.3.9)$$

where $n_\text{H} \equiv n_{\text{HI}} + n_{\text{HII}}$ is the total hydrogen density. Also, ignoring all elements but hydrogen, the electrical neutrality of the interstellar medium requires

$$n_e = n_{\text{HII}} = n_\text{H} - n_{\text{HI}}.$$

Hence for pure hydrogen the cooling function (3.3.4) takes the form

$$\Lambda = \Lambda_0 n_\text{H}^2 \exp(-7T_\text{I}/4T)\bigl(\beta(T)/\alpha(T)\bigr)\bigl(1 + \exp(-T_\text{I}/T)\beta(T)/\alpha(T)\bigr)^{-2}, \quad (3.3.10)$$

in which we have written the exponential in Eq. (3.3.4) as $\exp(-3T_\text{I}/4T)$. At the temperatures of interest $\beta(T)/\alpha(T)$ is very large and slowly varying, so the cooling function rises steeply for small temperature, proportionally to $\exp(-7T_\text{I}/4T)$, reaches a maximum at $\exp(-T_\text{I}/T) = 7\alpha/\beta$, and then falls off with increasing temperature, reaching a plateau value proportional to $\alpha/\beta \ll 1$. For instance, if we take β/α to have the value $\beta/\alpha = 5 \times 10^4$ that the estimates (3.3.6) and (3.3.8) give for pure hydrogen at $15{,}000\,\text{K}$, then the maximum in the cooling curve is at $T \simeq T_\text{I}/8.9 \simeq 17{,}700\,\text{K}$.

Including a typical admixture of helium, $n_{\text{He}} \simeq n_\text{H}/12$, gives a second local maximum in the cooling curve at $T \simeq 10^5\,\text{K}$, with lower height than the first maximum, and the cooling function then again falls off rapidly for greater

---
[18] Mo et al., op. cit., Eqs. (B.1.55) and (B.1.56).

temperature. The cooling is greatly enhanced if the interstellar medium is enriched by metals, such as carbon and oxygen and their ions, which have lower excitation energies.

We can understand the energy levels of multi-electron atoms and ions in the Hartree approximation, according to which each electron can be considered to move in an effective potential produced by the nucleus and all the other electrons. The effective potential is close to being spherically symmetric, so, apart from spin, the states of individual electrons can be characterized by an orbital angular momentum $\ell = 0, 1, 2$, etc. and a principal quantum number $n \geq \ell + 1$, defined so that the number of nodes of the radial wave function is $n - \ell - 1$. The effective potential is not of the Coulomb form, so one-particle states of the same n and different ℓ are generally not even approximately degenerate. Since the z-component of orbital angular momentum takes $2\ell + 1$ values, and the z-component of the electron spin takes 2 values, there are $2(2\ell + 1)$ one-electron states with a given n and ℓ, and this is the number of electrons that are allowed by the Pauli exclusion principle to occupy such a single-particle state. The two most tightly bound electrons will have $n = 1$ and $\ell = 0$, and the two next most tightly bound will have $n = 2$ and $\ell = 0$. Atoms with five to ten electrons, and in particular the most abundant metals, namely carbon, nitrogen, and oxygen, will have their least bound electrons in $n = 2$, $\ell = 1$ single-particle states. Even apart from relativistic and magnetic effects, the corrections to the Hartree approximation give the atom's energies a dependence on L, the total orbital angular momentum, with the energies of states with different L differing by a few eV. As we will see, in cases of interest here the requirement that the wave function be completely antisymmetric in electron variables implies that states with definite values of L also have definite values of the total electron spin S.

For states with definite non-zero L and S, comparatively weak magnetic and relativistic effects produce a small fine-structure splitting into states of definite energy, the higher of which can be excited even at quite low temperature. Rotational invariance requires these states to have definite values for the total angular momentum J. These states of definite energy, L, S, and J, are described in a "Russell–Saunders" notation, as

$$^{2S+1}L_J,$$

with the value of L indicated as S, P, D, F, etc. for $L = 0, L = 1, L = 2, L = 3$, etc.

For example, the singly ionized carbon atom, CII, has five electrons: the two that are most deeply bound in states with $n = 1$, $\ell = 0$; the two that are next most deeply bound in states with $n = 2$, $\ell = 0$; and a single least bound "valence" electron in a state with $n = 2$, $\ell = 1$. The coupling of the orbital and spin angular momenta of the valence electron splits this state into components with two different values for the total angular momentum, $J = 1/2$

and $J = 3/2$, denoted $^2P_{1/2}$ and $^2P_{3/2}$, split by an energy $k_B \times 92\,K$. The upper state with $J = 3/2$ can be excited by collisions with hydrogen atoms or molecules as well as with electrons, allowing cooling to temperatures of a few tens of degrees Kelvin.

Because oxygen is more abundant than any other metal, the case of doubly ionized oxygen, OIII, is of special interest. These ions have the two most deeply bound electrons in states with $n = 1$, $\ell = 0$; the two next most deeply bound electrons in states with $n = 2$, $\ell = 0$, and *two* valence electrons in states with $n = 2$, $\ell = 1$. These two $\ell = 1$ orbital angular momenta can add up to a total orbital angular momentum $L = 0$, $L = 1$, or $L = 2$. The states with $L = 0$ and $L = 2$ are symmetric in the orbital angular momentum quantum numbers of the two valence electrons, so since electrons are fermions the wave function must be antisymmetric in the spin quantum numbers of these two electrons, and hence the total spin is $S = 0$ and these states therefore have no fine structure. They are denoted $^1S_0$ and $^1D_2$. On the other hand, the state with total orbital angular momentum $L = 1$ is antisymmetric in the orbital angular momentum quantum numbers of the two valence electrons, so the wave function must be symmetric in spin quantum numbers of these two electrons, and hence the total spin is $S = 1$. The $L = 1$ states are therefore split into components $^3P_0$, $^3P_1$, and $^3P_2$ with total angular momentum $J = 0$, $J = 1$, and $J = 2$. As it happens, the lowest OIII states are those with $L = 1$; the states with $L = 2$ and $L = 0$ are higher by about 2.5 eV and 5.3 eV, respectively. Among these $L = 1$ states, the lowest is $^3P_0$, with $J = 0$; the $^3P_1$ and $^3P_2$ states with $J = 1$ and $J = 2$ are higher by energies 0.014 eV and 0.038 eV, respectively, corresponding to temperatures 162 K and 441 K, respectively, so collisions of electrons with OIII ions can provide a mechanism for cooling down to about 100 K. Similar remarks apply to the NII ion, which, however, is generally less abundant than OIII, and therefore less important in cooling.

One last example: The singly ionized OII ion has three $\ell = 1$ valence electrons. In the most deeply bound state these three unit orbital angular momenta add up to a total orbital angular momentum $L = 0$, for which the wave function is completely antisymmetric in valence electron coordinates. It therefore must be completely symmetric in valence electron spins, so this state has $S = 3/2$ and hence $J = 3/2$. With $L = 0$, this $^4S_{3/2}$ state of course has no fine-structure splitting. The first excited states are a $^2D_{3/2}$, $^2D_{5/2}$ doublet, about 3.3 eV above the ground state.

These ions provide the chief mechanism for cooling in ionized clouds, such as HII regions.[19] For electron temperatures below about 6,000 K, cooling is dominated by electronic excitation of the fine-structure transitions $^3P_0 \to {}^3P_1$ and $^3P_0 \to {}^3P_2$ in OIII. For higher temperatures cooling is dominated by

---

[19] Results in this paragraph are taken from Figure 6.1 of L. Spitzer, Jr., *Physical Processes in the Interstellar Medium* (John Wiley & Sons, New York, 1998),

electronic excitation of the 3.3 eV transitions $^4S_{3/2} \rightarrow {}^2D_{3/2}$ and $^4S_{3/2} \rightarrow {}^2D_{5/2}$ in OII. At these temperatures there is also significant but less important cooling from the $^3P \rightarrow {}^1D_2$ transitions in OIII and NII. Setting the cooling function calculated in this way equal to the heating function (3.2.21) gives an equilibrium electron temperature of about 8,000 K for an HII region heated by a star with color temperature 35,000 K.

Delayed Radiation

In some cases collisions are so rapid and radiative decays relatively so slow that collisional de-excitation is much faster than radiation, and the states excited in collisions usually lose their energy in further collisions, with the loss of energy by photon emission relatively rare. In this case the number density of atoms or molecules in an energy level with energy E and degeneracy g is fixed by conditions of thermal equilibrium to be approximately proportional to $g \exp(-E/k_B T)$, so the number density in an excited energy level with energy E_a and degeneracy g_a is related to the total number density $n = \sum_a n_a$ of these atoms or molecules in any state by

$$n_a = n Z^{-1}(T) g_a \exp\left(-[E_a - E_0]/k_B T\right), \qquad (3.3.11)$$

where

$$Z(T) \equiv \sum_a g_a \exp\left(-[E_a - E_0]/k_B T\right), \qquad (3.3.12)$$

and E_0 is conveniently taken as the lowest energy level. If A_b^a (in the notation introduced by Einstein) is the rate at which states of energy E_b (averaged over such states, if $g_b > 1$) spontaneously decay into any state with energy $E_a < E_b$ (including the lowest energy level $a = 0$) with the emission of radiation of energy $E_b - E_a$, then the rate of energy loss per time and per volume is

$$\begin{aligned}\Lambda &= \sum_{b:E_b>E_0} n_b \sum_{a:E_a<E_b} A_b^a(E_b - E_a) \\ &= nZ^{-1}(T) \sum_{b:E_b>E_0} g_b \exp\left(-[E_b - E_0]/k_B T\right) \sum_{a:E_a<E_b} A_b^a(E_b - E_a),\end{aligned}$$
$$(3.3.13)$$

so here the cooling function is proportional to the density. As long as we know that the collision rate is high eneough to keep target particles in thermal equilibrium and for the de-excitation transition $b \rightarrow a$ to be dominated by collisions rather than radiation, we do not need to know the collision rate, but we do need to know the radiative transition rate A_b^a to calculate the cooling function.

This case is commonly encountered for the most common molecule in interstellar space, diatomic hydrogen. In general, the lowest energy levels of molecules are rotational[20] – vibrational modes have energies higher by a factor of order $(m_p/m_e)^{1/2}$, and modes with electronic excitation have energies still higher, by a factor of order m_p/m_e. At sufficiently low temperature the hydrogen molecule can therefore be treated as a rigid linear rotator, whose only continuous degree of freedom is the direction $\hat{n}$ of the separation of the two protons. The wave functions are spherical harmonics, $Y_\ell^m(\hat{n})$, with $\ell = 0, 1, 2, \ldots$ and m an integer taking $2\ell + 1$ values, from $-\ell$ to $+\ell$. The energies of the corresponding states are of the form $E_\ell = \hbar^2 \ell(\ell+1)/2I$, where I is the moment of inertia, which for hydrogen molecules takes a value for which $\hbar^2/2Ik_B = 45$ K.

There is an important complication here, due to the fact that protons are fermions, with spin 1/2. The overall wave function must be antisymmetric in exchange of the spin and coordinates of the two protons, so the states of total spin zero and one, known as parahydrogen and orthohydrogen, which are respectively antisymmetric and symmetric in proton spins, must be respectively even or odd in $\hat{n}$, and therefore have ℓ even or odd, respectively. Aside from this limitation on ℓ, the alignment of spins has a negligible effect on energy, so that both orthohydrogen and parahydrogen molecules have energies $E_\ell = \hbar^2 \ell(\ell+1)/2I$. For total spin s the 3-component of the spin takes $2s + 1$ values, so the degeneracies of hydrogen molecules are

$$g_{\ell,\text{para}} = 2\ell + 1, \qquad g_{\ell,\text{ortho}} = 3(2\ell + 1). \tag{3.3.14}$$

The lowest energy level of orthohydrogen is higher than the lowest energy level of parahydrogen by an energy $2 \times \hbar^2/2I = 90\,\text{K}/k_B$, so at temperatures well above about 100 K there are three orthohydrogen molecules for each parahydrogen molecule.

The rate for radiative conversion between orthohydrogen and parahydgrogen is extremely low, so radiative transitions in hydrogen molecules take place only between states which both have ℓ even, or both have ℓ odd, and therefore do not change the parity of the molecule. The dominant radiative transitions are therefore those in which ℓ changes by two units, and Eq. (3.3.13) gives

$$\Lambda = n_{H_2} Z^{-1}(T) \left[\sum_{\ell \geq 2 \text{ even}} + 3 \sum_{\ell \geq 3 \text{ odd}} \right]$$
$$\times (2\ell + 1) \exp\left(-\hbar^2 \ell(\ell+1)/2Ik_B T\right) A_\ell^{\ell-2} (4\ell - 2)\hbar^2/2I, \tag{3.3.15}$$

where

$$Z(T) = \left[\sum_{\ell \geq 0 \text{ even}} + 3 \sum_{\ell \geq 1 \text{ odd}} \right] (2\ell + 1) \exp\left(-\hbar^2 \ell(\ell+1)/2Ik_B T\right).$$

---

[20] For a textbook treatment of rotational and vibrational modes of molecules, see S. Weinberg, *Lectures on Quantum Mechanics*, 2nd edn. (Cambridge University Press, Cambridge, 2015), Sections 4.9 and 5.6.

This gives a cooling function at 100 K equal to $n_{H_2} \times 3 \times 10^{-26}$ erg/sec.[21] Radiative de-excitation is slower in H_2 molecules than collisional de-excitation because, as we have noted, the dominant radiative transitions in H_2 are between states in which ℓ changes by two units. These are electric quadrupole transitions, and therefore slower than electric dipole transitions by a factor of order $(a/\lambda)^2$, where a is the length of the molecule and λ is the wavelength of the emitted photon.

Radiative transitions are faster in molecules such as HD or CO, in which the nuclei are not identical. For such molecules electric dipole transitions between states with ℓ differing by one unit are allowed, and are typically faster than collisional de-excitation. Thus, although such molecules are always much less abundant than H_2, they can make a larger contribution to the cooling function. This is especially the case at low temperatures for CO, which has a much larger moment of inertia than H_2 or HD, and therefore has a very low excitation energy of just 5.5 K $\times\, k_B$ for the transition $\ell = 0 \to \ell = 1$. In very cold clouds in the galactic plane it is CO that provides the dominant cooling.

Bremsstrahlung Cooling

As we saw earlier in this section, at temperatures above about 15,000 K hydrogen is mostly ionized, and cooling can no longer occur through the excitation of neutral hydrogen. Instead, at these temperatures cooling occurs through bremsstrahlung, the radiation of photons in the Coulomb scattering of electrons by ions. This is the inverse of the process leading to free–free absorption of photons, discussed in Section 1.4.

In order to calculate the contribution of bremsstrahlung to the cooling function, we need to know the bremsstrahlung emissivity $j_\nu(T)$, the rate of emission of radiation energy per volume, per solid angle, and per frequency interval at temperature T of emission of photons of frequency ν, given by

$$j_\nu(T) = h\nu \int \frac{d^3q}{4\pi} \delta(qc/h - \nu) \int d^3p \, n_e(\mathbf{p}, T) \sum_\lambda \int d^3p' \, j(\mathbf{q}, \lambda; \mathbf{p} \to \mathbf{p}'), \tag{3.3.16}$$

where $n_e(\mathbf{p}, T) \, d^3p$ is the number density of electrons in a range d^3p of momenta around initial momentum $\mathbf{p}$ at temperature T, and $j(\mathbf{q}, \lambda; \mathbf{p} \to \mathbf{p}') \, d^3p'$ is the rate at which an electron of initial momentum $\mathbf{p}$ is scattered into a range d^3p' of momenta around a final momentum $\mathbf{p}'$, with the emission of a photon with helicity λ and with momentum in a range d^3q around $\mathbf{q}$. According to the general rules of quantum mechanics, the rate per electron is given by

---

[21] J. E. Dyson and D. A. Williams, *The Physics of the Interstellar Medium*, 2nd edn. (Institute of Physics Publishing, Bristol, 1997), p. 31.

$$j(\mathbf{q}, \lambda, \mathbf{p} \to \mathbf{p}') = (2\pi\hbar)^2 n_N |M|^2 \delta\left(\frac{p^2}{2m_e} - \frac{p'^2}{2m_e} - qc\right), \qquad (3.3.17)$$

where n_N is the number density of nuclei, and the matrix element M is the coefficient of the energy-momentum delta function in the S-matrix for the process in which an electron with momentum $\mathbf{p}$ and a nucleus at rest yields a photon with momentum $\mathbf{q}$ and helicity λ, an electron with momentum $\mathbf{p}'$, and a recoiling nucleus. (A momentum conservation delta function is implicitly used here to fix the recoil momentum as $\mathbf{p} - \mathbf{p}' - \mathbf{q}$. Since the nucleus is relatively so massive, with this momentum it carries off negligible kinetic energy.) To first order in the Coulomb potential, M is the same as the Born-approximation matrix element (1.4.A10) for the corresponding absorption process, except that the polarization vector $e(\mathbf{q}, \lambda)$ is replaced with its complex conjugate:

$$M^{\text{Born}} = \frac{Ze^3}{2\pi^2 (qc\hbar)^{3/2} m_e} \frac{\mathbf{e}^*(\hat{q}, \lambda) \cdot (\mathbf{p} - \mathbf{p}')}{(\mathbf{p} - \mathbf{p}')^2}, \qquad (3.3.18)$$

where Ze is the nuclear charge (again in unrationalized electrostatic units). Using this in Eq. (3.3.17) gives the emissivity per electron

$$j^{\text{Born}}(\mathbf{q}, \lambda; \mathbf{p} \to \mathbf{p}') = \frac{Z^2 e^6 n_N}{\pi^2 \hbar [(\mathbf{p}-\mathbf{p}')^2]^2 q^3 c^3 m_e^2}$$
$$\times \left|(\mathbf{p}-\mathbf{p}') \cdot \mathbf{e}^*(\lambda, \mathbf{q})\right|^2 \delta\left(\frac{p^2}{2m_e} - \frac{p'^2}{2m_e} - qc\right). \qquad (3.3.19)$$

Using Eq. (1.4.A12), we see that the average of this over photon direction and summed over helicity is

$$\sum_\lambda \int \frac{d^2\hat{q}}{4\pi} j^{\text{Born}}(\mathbf{q}, \lambda; \mathbf{p} \to \mathbf{p}') = \frac{2Z^2 e^6 n_N}{3\pi^2 \hbar (\mathbf{p}-\mathbf{p}')^2 q^3 c^3 m_e^2}$$
$$\times \delta\left(\frac{p^2}{2m_e} - \frac{p'^2}{2m_e} - qc\right). \qquad (3.3.20)$$

The integral over the directions of $\mathbf{p}'$ is given by Eq. (1.4.A13)

$$\int \frac{d^2\hat{p}'}{|\mathbf{p}-\mathbf{p}'|^2} = \frac{2\pi}{pp'} \ln\left(\frac{p+p'}{p-p'}\right),$$

and we find the Born-approximation bremsstrahlung emissivity:

$$j_\nu(T) = \frac{4Z^2 e^6 n_N}{3\pi \hbar c^3 m_e^2} \int n_e(\mathbf{p}, T) d^3p$$
$$\times \int_0^p \frac{p'^2 dp'}{pp'} \ln\left(\frac{p+p'}{p-p'}\right) \delta\left(\frac{p^2}{2m_e h} - \frac{p'^2}{2m_e h} - \nu\right). \qquad (3.3.21)$$

Assuming that all photons leave the medium, the contribution of bremsstrahlung to the rate of energy loss per volume and per time is given in the Born approximation by

$$\Lambda_{\text{brem}}(T) = 4\pi \int j_\nu(T)\, d\nu. \tag{3.3.22}$$

The integral over ν is trivial, and we have

$$\Lambda_{\text{brem}}(T) = \frac{16 Z^2 e^6 n_N}{3\hbar c^3 m_e^2} \int n_e(\mathbf{p})\, d^3 p \int_0^p \frac{p'^2\, dp'}{pp'} \ln\left(\frac{p+p'}{p-p'}\right).$$

Now the integral over p' is a standard definite integral:

$$\int_0^p p'\, dp'\, \ln\left(\frac{p+p'}{p-p'}\right) = p^2.$$

The bremsstrahlung cooling function (3.3.22) is then

$$\Lambda_{\text{brem}}(T) = \frac{16 Z^2 e^6 n_N}{3\hbar c^3 m_e^2} \int p\, n_e(\mathbf{p}, T)\, d^3 p. \tag{3.3.23}$$

If electrons are in thermal equilibrium at temperature T, then $n_e(\mathbf{p})$ is given by the Maxwell–Boltzmann momentum distribution

$$n_e(\mathbf{p}, T) = (2\pi m_e k_B T)^{-3/2} n_e \exp\left(-p^2 / 2 m_e k_B T\right), \tag{3.3.24}$$

where n_e here is the total number density of electrons. Then the integral over p is

$$\int p\, n_e(\mathbf{p}, T)\, d^3 p = n_e \sqrt{8 m_e k_B T / \pi},$$

and the contribution of bremsstrahlung to the cooling function in the Born approximation is

$$\Lambda_{\text{brem}}(T) = \frac{16 Z^2 e^6 n_N n_e}{3\pi \hbar c^3 m_e^2} \sqrt{8\pi m_e k_B T}. \tag{3.3.25}$$

Although this result is approximate, the growth of the bremsstrahlung cooling function as $\sqrt{T}$ is generally seen in more elaborate calculations.

To summarize results for cooling: Eq. (3.3.10) shows that in pure hydrogen the cooling function rises rapidly at first with increasing temperature, as more and more hydrogen atoms become energetic enough to excite the $2s$ state and higher states in collisions, and then for $T \simeq 15{,}000\,\text{K}$ begins a rapid decline with increasing temperature as more and more of the hydrogen is ionized. The presence of metals in the form of molecules, ions, or atoms then dominates the cooling. For pure hydrogen, at about $10^6\,\text{K}$ the decline of the cooling function is replaced with a slow rise of Λ with increasing temperature, as bremsstrahlung becomes the chief mechanism for cooling the ionized hydrogen.[22]

3.4 Star Formation

We saw in Section 1.1 that an isolated spherical mass in hydrostatic equilibrium obeys the virial theorem
$$\Pi + \Omega = 0, \tag{3.4.1}$$
where Π is a measure of the thermal energy
$$\Pi \equiv 3 \int_0^R 4\pi r^2 p(r)\, dr, \tag{3.4.2}$$
and Ω is the gravitational self-energy
$$\Omega = -G \int_0^R 4\pi r \rho(r) \mathcal{M}(r)\, dr, \tag{3.4.3}$$
where
$$\mathcal{M}(r) \equiv \int_0^r 4\pi r'^2 \rho(r')\, dr'. \tag{3.4.4}$$

We can think of Eq. (3.4.1) as marking a boundary between two kinds of non-equilibrium configurations. If Π is greater than $-\Omega$ then the gravitational field is not strong enough to prevent pressure forces from dispersing the matter. On the other hand, if Π is less than $-\Omega$, then the pressure is not strong enough to prevent the beginning of gravitational condensation into a star.

Of course, without an equilibrium condition such as Eq. (1.1.4), there is nothing to pick out any particular initial distribution of density and pressure in the mass. But we can make a crude estimate. The density $\rho(r)$ may be taken to be of the order of a typical value $\bar{\rho}$ of density; M and $\mathcal{M}(r)$ are of order $4\pi \bar{\rho} R^3$; and $p(r)$ is of the order $c_s^2 \bar{\rho}$, where c_s is of the same order as the typical acoustic velocity. Then, ignoring all dimensionless factors of order unity other than 4π,
$$\Pi \approx 4\pi c_s^2 \bar{\rho} R^3, \qquad -\Omega \approx (4\pi)^2 G \bar{\rho}^2 R^5. \tag{3.4.5}$$
Hence the condition $\Pi < |\Omega|$ for gravitational condensation is, roughly,
$$R > R_{\rm J}, \tag{3.4.6}$$

---

[22] The cooling curves for hydrogen plus helium, with and without metals, are displayed in Figure 8.1 of H. Mo, F. van den Bosch, and S. White, *Galaxy Formation and Evolution* (Cambridge University Press, Cambridge, 2010).

where R_J is the so-called *Jeans length*:

$$R_J^2 \equiv \frac{c_s^2}{4\pi G \bar{\rho}}. \tag{3.4.7}$$

The velocity c_s can be estimated to have the adiabatic value $c_s^2 = \Gamma k_B T/m$, where Γ is 5/3 or 7/5 for monatomic and diatomic gases, respectively, and the particle mass m has the value Am_p where m_p is the proton mass, and A is the atomic (or molecular) weight in units of m_p. Then the Jeans length is given by

$$R_J^2 \equiv \frac{\Gamma k_B T}{4\pi G A m_p \bar{\rho}}. \tag{3.4.8}$$

This calculation only gives an order-of-magnitude estimate for the scale above which gravitational condensation occurs, but we will see below that with our result for R_J, $q_J = 1/R_J$ is precisely the critical wave number (though in a calculation based on unrealistic assumptions) at which there is a transition from stable oscillation to instability.

Since the linear scale and density of a condensation change during collapse, it is convenient to work with an invariant, the Jeans mass, equal to the mass within a sphere of diameter equal to the corresponding wavelength $2\pi/q_J = 2\pi R_J$:

$$M_J \equiv \frac{4\pi}{3}\bar{\rho}(\pi R_J)^3 = \frac{\pi^{5/2} c_s^3}{6 A^{1/2} m_p^{1/2} n^{1/2} G^{3/2}}$$

$$= 6.57 \times 10^4 M_\odot (c_s \, [\text{km/sec}])^3 (n \, [\text{cm}^{-3}])^{-1/2} A^{-1/2}, \tag{3.4.9}$$

where $n = \bar{\rho}/Am_p$ is the number density. Setting $c_s^2 = \Gamma k_B T/m_p A$, this is

$$M_J = \frac{\pi^{5/2}(\Gamma k_B T)^{3/2}}{6 A^2 m_p^2 n^{1/2} G^{3/2}} = 49.5 M_\odot A^{-2} (\Gamma T \, [\text{K}])^{3/2} (n \, [\text{cm}^{-3}])^{-1/2}. \tag{3.4.10}$$

Stars are being formed in the plane of our galaxy in giant molecular clouds, with cloud masses extending up to about $6 \times 10^6 M_\odot$. It is important that these clouds are rich in molecules, such as H_2 and CO, because as we saw in Section 3.3 atomic hydrogen is ineffective in cooling at temperatures below about 15,000 K because of the high excitation energy of the first $n = 2$ excited state, and ineffective in cooling at temperatures above about 15,000 K because it is mostly ionized. On the other hand, molecules (and also ions such as OIII) have much lower excited states, which can be excited at much lower temperatures and lose their energy through radiation. (Ultraviolet light that would otherwise dissociate the molecules in molecular clouds is usually blocked by surrounding atomic or molecular hydrogen.) Cooling is essential in star formation, because it lowers c_s, and hence lowers the Jeans mass.

For a giant molecular cloud with number density $n = 10^4$ cm^{-3}, temperature $T = 20$ K, molecular weight $A = 2$, and adiabatic index $\Gamma = 7/5$, Eq. (3.4.10)

3.4 Star Formation

gives a Jeans mass of $18 M_\odot$. This is not an unreasonable mass for a protostar, which suffers further fragmentation into stars like the Sun. As the protostar shrinks, its density increases, which lowers the Jeans mass. However, if its temperature also increases, that increases the Jeans mass. The fragmentation into smaller stars therefore requires an efficient cooling mechanism. Another complication is turbulence, which can play the role here of the acoustic velocity c_s. Taking the turbulent velocities to be of order 1 to 10 km/sec and $A = 2$, then even for n as large as 10^5 cm^{-3}, Eq. (3.4.9) gives a Jeans mass of order 150 to 150,000 solar masses. On the other hand, star formation can be assisted by the compression of the interstellar medium by supernova shock waves or stellar winds. This is a complicated business, too complicated for detailed treatment here.[23]

The radius (3.4.7) is called the Jeans length because in 1902 James Jeans for the first time gave a value for the minimum wavelength of density perturbations that grow exponentially, which turned out to be of order (3.4.7). Though apparently precise, as we shall see, Jeans' calculation was in fact no more accurate than the crude estimate given above. His work was based on the Newtonian equations of fluid dynamics: the continuity equation for the density and the fluid velocity $\mathbf{u}$,

$$\frac{\partial \rho}{\partial t} + \nabla \cdot (\mathbf{u}\rho) = 0; \tag{3.4.11}$$

Euler's equation (essentially Newton's second law),

$$\frac{\partial \mathbf{u}}{\partial t} + (\mathbf{u} \cdot \nabla)\mathbf{u} = -\nabla \phi - \frac{1}{\rho}\nabla p; \tag{3.4.12}$$

and the Poisson equation for the gravitational potential ϕ,

$$\nabla^2 \phi = 4\pi G \rho. \tag{3.4.13}$$

The solution is to be written as a set of assumed unperturbed quantities, labeled with a subscript 0, plus small perturbations to be calculated, labeled with a subscript 1:

$$\rho = \rho_0 + \rho_1, \quad \mathbf{u} = \mathbf{u}_0 + \mathbf{u}_1, \quad \text{etc.} \tag{3.4.14}$$

Jeans took the unperturbed velocity to vanish, and all the other unperturbed quantities to be constants in both space and time. The terms in Eqs. (3.4.11)–(3.4.13) of first order in perturbations are then

$$\frac{\partial \rho_1}{\partial t} + \rho_0 \nabla \cdot (\mathbf{u_1}) = 0, \tag{3.4.15}$$

$$\frac{\partial \mathbf{u}_1}{\partial t} = -\nabla \phi_1 - \frac{c_s^2}{\rho_0} \nabla \rho_1, \tag{3.4.16}$$

---

[23] For a taste of these complications, see C. F. McKee and E. C. Ostriker, *Ann. Rev. Astron. Astrophys.* **45**, 1 (2007).

and
$$\nabla^2 \phi_1 = 4\pi G \rho_1, \qquad (3.4.17)$$

where
$$c_s^2 \equiv \left(\frac{\partial p}{\partial \rho}\right)_{\rho=\rho_0}, \qquad (3.4.18)$$

the derivative being taken with constant entropy per baryon or constant temperature, as appropriate in the circumstances. Since the unperturbed background is constant in space and time, we can find a solution in which the spacetime dependence of all perturbations is given by a factor $\exp(-i\omega t + i\mathbf{q} \cdot \mathbf{x})$, with ω and $\mathbf{q}$ constant. Equations (3.4.17) and (3.4.16) then give in turn

$$\phi_1 = -\frac{4\pi G}{q^2} \rho_1$$

and
$$\mathbf{u}_1 = \frac{\mathbf{q}}{\omega}\left(\phi_1 + \frac{c_s^2}{\rho_0}\rho_1\right) = \frac{\mathbf{q}\rho_1}{\omega}\left(-\frac{4\pi G}{q^2} + \frac{c_s^2}{\rho_0}\right),$$

and using this in Eq. (3.4.15) then gives

$$i\omega\rho_1 = i\mathbf{q} \cdot \mathbf{u}_1 \rho_0 = \frac{iq^2\rho_1}{\omega}\left(-\frac{4\pi G\rho_0}{q^2} + c_s^2\right),$$

or, canceling factors of $i\rho_1/\omega$,

$$\omega^2 = c_s^2 q^2 - 4\pi G \rho_0. \qquad (3.4.19)$$

This result may be usefully compared with the corresponding result for electromagnetic waves in a plasma

$$\omega^2 = c_s^2 q^2 + 4\pi e^2 n_e,$$

where n_e is the number density of electrons. For large wave numbers q, both relations reduce to the usual formula $\omega = \pm c_s q$ for sound waves, but for small q there is a very large difference: the perturbations described by Eq. (3.4.19) grow exponentially with time for q^2 less than $4\pi G\rho_0/c_s^2$, or in other words, for $1/q$ greater than the Jeans length R_J, while electromagnetic waves do nothing but oscillate at any wave number. The crucial difference of sign in the second term of Eq. (3.4.19) is of course due to the fact that electrical forces between electrons are repulsive, while gravitational forces between all particles are attractive.

We should not conclude from the derivation of Eq. (3.4.19) that there really is a sharp transition at wave number $1/R_J$ between oscillation and exponential growth of perturbations. The trouble is that the assumed unperturbed "solution" is not a solution at all. Equation (3.4.13) shows that if ρ is constant, then ϕ cannot be constant. This is not just a point of principle. If we take the unperturbed solution to be time-independent then the equilibrium condition (1.1.4) shows

3.4 Star Formation

that the unperturbed fluid must vary over a scale of distances of order R_J, so we cannot estimate what happens at wave numbers of order $1/R_J$ by taking the unperturbed fluid to be constant in time and space.

We can, however, use Eq. (3.4.19) to estimate that in the limit of small wave numbers, the rate of growth of gravitational condensation is of order $\sqrt{4\pi G\rho}$. This can be verified without the use of perturbation theory. Let us consider an actual solution of Eqs. (3.4.11)–(3.4.13) for which the density ρ and pressure p are constant in space, though not in time. For simplicity, let us also assume that this solution is spherically symmetric around some point $r = 0$. With ρ constant in space, there is a spherically symmetric solution of Eq. (3.4.13):

$$\phi(r,t) = 2\pi G\rho(t)r^2/3. \qquad (3.4.20)$$

Then, with p constant in space, Eq. (3.4.12) has the solution

$$\mathbf{u} = H(t)\mathbf{r}, \quad \dot{H}(t) + H(t)^2 = -4\pi G\rho(t)/3, \qquad (3.4.21)$$

and Eq. (3.4.11) then gives

$$\rho(t) \propto a(t)^{-3}, \quad \dot{a}(t)/a(t) = H(t). \qquad (3.4.22)$$

We can write the differential equation (3.4.21) as a conservation law. Note that

$$\frac{d}{dt}(H^2 a^2 - 8\pi G\rho a^2/3) = 2Ha^2(\dot{H} + H^2 + 4\pi G\rho/3) = 0,$$

so

$$H^2 - 8\pi G\rho/3 = -K/a^2, \qquad (3.4.23)$$

with K constant in time as well as space. This of course is a non-relativistic version of the Friedmann model of an expanding (or contracting) universe. In cosmology, K is a measure of spatial curvature; here it is merely a constant of the motion.

We can use Eq. (3.4.23) to calculate the time taken for a cloud to collapse to a much smaller scale, where pressure effects may begin to be important. Suppose that the matter of the cloud is at rest, so that H vanishes, at some initial time t_0. Then according to Eq. (3.4.23) we have $K/a^2(t_0) = 8\pi G\rho(t_0)/3$, and Eq. (3.4.21) becomes

$$\left(\frac{\dot{a}(t)}{a(t)}\right)^2 = \frac{8\pi G\rho(t_0)}{3}\left[\left(\frac{a(t_0)}{a(t)}\right)^3 - \left(\frac{a(t_0)}{a(t)}\right)^2\right]. \qquad (3.4.24)$$

The time required for the collapse in free fall of the distance scale $a(t)$ from $a(t_0)$ to a much smaller value is then

$$T_c = \sqrt{\frac{3}{8\pi G\rho(t_0)}} \int_0^{a(t_0)} \frac{da}{a}\left[\left(\frac{a(t_0)}{a}\right)^3 - \left(\frac{a(t_0)}{a}\right)^2\right]^{-1/2} = \frac{\pi}{2}\sqrt{\frac{3}{8\pi G\rho(t_0)}}, \qquad (3.4.25)$$

in agreement with our earlier estimate of a growth rate of order $\sqrt{4\pi G \rho_0}$. Again setting $\rho(t_0) = n(t_0) m_\mathrm{p} A$, this is

$$T_\mathrm{c} = 51.5 \times 10^6 \text{ years} \times \left(An(t_0) \text{ [cm}^{-3}\text{]} \right)^{-1/2}. \tag{3.4.26}$$

This is generally a short time compared with typical galactic time scales.

3.5 Accretion Disks

Not all interstellar matter remains interstellar. Here and there in the plane of our galaxy interstellar matter is falling onto a neighboring star. The interstellar matter usually starts with enough angular momentum that the incoming matter speeds up as it falls down, and forms an accretion disk, revolving about the star as its inner parts fall onto the star's surface. It was from such a disk about the young Sun that the planets of the solar system were formed. More spectacular are the accretion disks around black holes at the centers of quasars, which will be discussed in Section 4.5. Although these various accretion disks vary enormously in scale, they can all be described using a common mathematical formalism, the subject of this section.

The great luminosity of many accretion disks is explained in part by the fact that they are not subject to a well-known limit on the luminosity that can be produced by spherically symmetric accretion onto a given mass. This is the *Eddington limit*, which we have already encountered in Eq. (1.3.17). To recapitulate, a spherically symmetric source of luminosity L produces a force per gram on ionized accreting matter at distance r equal to $(L/4\pi r^2 c)\kappa_\mathrm{T}$, where κ_T is the opacity due to Thomson scattering, equal to the Thomson scattering cross section times the number of free electrons per gram. For accretion to continue on a mass M, this cannot exceed the gravitational force per gram GM/r^2, so L cannot exceed the Eddington luminosity $L_\mathrm{E} \equiv 4\pi c G M/\kappa_\mathrm{T}$. For completely ionized hydrogen, $L_\mathrm{E} = 3.25 \times 10^4 L_\odot (M/M_\odot)$.

But accretion can be much more effective than this.[24] Consider accreting matter in a configuration with cylindrical rather than spherical symmetry, surrounding a central mass M, with the accreting matter in circular orbits around the axis of symmetry. Here the maximum luminosity L_max is set by the condition that at every point the repulsive force due to radiation pressure and centrifugal force should be no greater than gravitational attraction. The centrifugal force per mass is $\Omega^2(R)\,\mathbf{R}$, where $\Omega(R)$ is the angular frequency of bodies in an orbit of radius R, and $\mathbf{R}$ is a vector of magnitude R in the plane of the orbit pointing outward away from the axis of symmetry. That is, in Cartesian coordinates with symmetry around the z-axis, $\mathbf{R} = (x, y, 0)$. The gravitational force per mass

---

[24] This argument is given by S. Kato, J. Fukue, and S. Mineshige, *Black Hole Accretion Disks* (Kyoto University Press, Kyoto, 2008). They attribute it to a communication from M. A. Abramowicz.

is $-\nabla\phi$, where ϕ is the gravitational potential. The force per mass exerted by radiation is $\kappa_T \boldsymbol{\ell}/c$, where $\boldsymbol{\ell}$ is a vector pointing in the direction of the radiation flow whose magnitude is the energy per time and per area passing through a small area normal to $\boldsymbol{\ell}$. (For instance, for a source that emits radiation of luminosity L with spherical symmetry, we have $\boldsymbol{\ell} = (L/4\pi r^2)\hat{\mathbf{r}}$, where $\mathbf{r} = r\hat{\mathbf{r}}$ is the vector from the source. But here we are not assuming spherical symmetry!) Hence the maximum luminosity is set by the condition that at every point

$$\Omega^2(R)\mathbf{R} + \kappa_T \boldsymbol{\ell}/c \le \nabla\phi, \qquad (3.5.1)$$

the inequality applying to each component of the vectors. This sets a limit on the luminosity

$$L = \int_A \boldsymbol{\ell}\cdot d\mathbf{A} \le \frac{c}{\kappa_T}\int_A \left[-\Omega^2(R)\mathbf{R} + \nabla\phi\right]\cdot d\mathbf{A}, \qquad (3.5.2)$$

where the integrals are over the surface of a large volume enclosing the region of accretion, and $d\mathbf{A}$ is a vector normal to this surface with magnitude equal to an element of the surface area. To calculate the surface integrals on the right, we use Gauss's theorem, so that the maximum luminosity is

$$L_{\max} = \frac{c}{\kappa_T}\int_V \left[-\nabla\cdot[\Omega^2(R)\mathbf{R}] + \nabla^2\phi\right]dV, \qquad (3.5.3)$$

the integral now taken over the whole accreting region. The Poisson equation gives $\nabla^2\phi = 4\pi G\rho$, and since $\nabla\cdot\mathbf{R} = 2$, we have

$$\nabla\cdot[\Omega^2(R)\mathbf{R}] = 2\Omega^2(R) + R\frac{d}{dR}\Omega^2(R).$$

Putting this together, the maximum luminosity is

$$L_{\max} = L_E - \frac{c}{\kappa_T}\int_V \left[2\Omega^2(R) + R\frac{d}{dR}\Omega^2(R)\right]dV, \qquad (3.5.4)$$

with $L_E = 4\pi GMc/\kappa_T$ the Eddington luminosity. The luminosity thus can exceed the Eddington limit if and only if the integral in Eq. (3.5.4) is negative. For rigid rotation, $\Omega(R)$ is independent of R, and this integral is positive. On the other hand, with the mass of the system concentrated in a central body, we expect Keplerian orbits, with $\Omega(R) \propto R^{-3/2}$, in which case Eq. (3.5.4) gives a maximum luminosity

$$L_{\max} = L_E + \frac{c}{\kappa_T}\int_V \Omega^2(R)\,dV, \qquad (3.5.5)$$

greater than the Eddington limit.

But if matter is in orbit around a central mass, how can it be accreted? After all, the planets of the solar system are not being pulled into the Sun. It is essential that here we are not dealing with a collisionless medium, like the solar system, but rather with a viscous gas in differential rotation. The viscosity here may arise from momentum transfer not so much by particles of long mean free path, as from effects of magnetic fields, such as Alfvén waves. But it can approximately be treated like an ordinary gas viscosity, and we will do so here.

Consider any circle around the axis of cylindrical symmetry. With $R\Omega$ a decreasing function of R, as for Keplerian orbits, the matter a little outside the circle is moving a little slower than the matter a little inside the circle. In the presence of viscosity, this shear causes the inside matter to slow down and hence move toward the center. (The implications of this simple picture for the flow of angular momentum will be discussed later.)

So we need to consider the effect of viscosity in gas rotating around an axis of symmetry. The dynamics of a non-relativistic viscous fluid in a gravitational potential ϕ is governed by the equation for the rate of change of momentum, the Navier–Stokes equation[25]

$$\frac{\partial}{\partial t}(\rho v_i) + \frac{\partial}{\partial x_j}(\rho v_i v_j) = -\rho \frac{\partial \phi}{\partial x_i} - \frac{\partial p}{\partial x_i} \\ + \frac{\partial}{\partial x_j}\left[\rho \nu \left(\frac{\partial v_i}{\partial x_j} + \frac{\partial v_j}{\partial x_i} - \frac{2}{3}\delta_{ij}\left(\frac{\partial v_k}{\partial x_k}\right)\right) \\ + \zeta \delta_{ij}\left(\frac{\partial v_k}{\partial x_k}\right)\right], \quad (3.5.6)$$

and the equation of mass conservation

$$\frac{\partial \rho}{\partial t} + \frac{\partial}{\partial x_i}(\rho v_i) = 0. \quad (3.5.7)$$

Here i, j, k are Cartesian vector indices, running over the values $1, 2, 3$; these indices when repeated are summed; ρ is the mass density; v_i are the Cartesian velocity components; p is the pressure; and ν and ζ are coefficients characterizing the fluid, known respectively as the *kinematic viscosity* and the *bulk viscosity*.[26]

For our present purposes, we need to use cylindrical coordinates, R, θ, z, and assume cylindrical symmetry, so that nothing depends on θ. Then the θ-component of Eq. (3.5.6) becomes

---

[25] For a classic textbook account, see L. D. Landau and E. M. Lifshitz, *Fluid Mechanics*, trans. J. B. Sykes and W. H. Reid (Pergamon Press, London, 1959).

[26] Since the contribution of bulk viscosity takes the same form as that of pressure, in defining bulk viscosity it is necessary to be careful about the definition of pressure. Usually one defines the pressure to be the same function of density and temperature as for uniform fluid flow, for which the viscosity terms in Eq. (3.5.6) would be absent. With that definition, the bulk viscosity vanishes for a non-relativistic or highly relativistic ideal gas, though it can be significant in a mixture of these. On this point, see S. Weinberg, *Astrophys. J.* **168**, 175 (1971). As a consequence of cylindrical symmetry, in the calculation below neither pressure nor bulk viscosity will play a role.

3.5 Accretion Disks

$$\frac{\partial}{\partial t}(\rho v_\theta) + \frac{1}{R^2}\frac{\partial}{\partial R}(R^2 \rho v_R v_\theta) + \frac{\partial}{\partial z}(\rho v_z v_\theta)$$
$$= \frac{1}{R^2}\frac{\partial}{\partial R}\left(R^3 \rho v \frac{\partial}{\partial R}\frac{v_\theta}{R}\right) + \frac{\partial}{\partial z}\left(\rho v \frac{\partial v_\theta}{\partial z}\right), \qquad (3.5.8)$$

and Eq. (3.5.7) reads

$$\frac{\partial \rho}{\partial t} + \frac{1}{R}\frac{\partial}{\partial R}(R\rho v_R) + \frac{\partial}{\partial z}(\rho v_z) = 0. \qquad (3.5.9)$$

We will limit our discussion here to a thin disk, concentrated around the plane $z=0$, in which ρ vanishes rapidly for large $|z|$, and within which nothing else varies appreciably with z. (The relation between ρ and z is discussed below.) Integrating Eqs. (3.5.8) and (3.5.9) over all z then gives

$$\frac{\partial}{\partial t}(\Sigma v_\theta) + \frac{1}{R^2}\frac{\partial}{\partial R}(R^2 \Sigma v_R v_\theta) = \frac{1}{R^2}\frac{\partial}{\partial R}\left(R^3 \Sigma v \frac{\partial}{\partial R}\frac{v_\theta}{R}\right) \qquad (3.5.10)$$

and

$$\frac{\partial \Sigma}{\partial t} + \frac{1}{R}\frac{\partial}{\partial R}(R\Sigma v_R) = 0, \qquad (3.5.11)$$

where Σ is the surface mass density

$$\Sigma(R,t) \equiv \int dz\, \rho(R,z,t). \qquad (3.5.12)$$

We next use Eqs. (3.5.10) and (3.5.11) to derive a differential equation relating the t and R dependence of the surface density (3.5.12). For this purpose, we assume that v_θ can be taken to be a time-independent function only of R. It is true that as gas moves inward or outward, the angular velocity of any element of gas changes at about the same fractional rate as Σ, but at any fixed R the value of v_θ depends only on R and $\phi'(R)$, and as long as the central mass is much larger than the mass in the disk the gravitational potential changes only very slowly. So we can take v_θ outside the time derivative in Eq. (3.5.10). We can also take the R derivative in the second term of Eq. (3.5.10) to act first on $R\Sigma v_R$ and then on Rv_θ, so that Eq. (3.5.10) becomes

$$v_\theta \frac{\partial}{\partial t}(\Sigma) + \frac{v_\theta}{R}\frac{\partial}{\partial R}(R\Sigma v_R) + \frac{\Sigma v_R}{R}\frac{\partial}{\partial R}(Rv_\theta) = \frac{1}{R^2}\frac{\partial}{\partial R}\left(R^3 \Sigma v \frac{\partial}{\partial R}\frac{v_\theta}{R}\right).$$

The first two terms cancel according to Eq. (3.5.11), so

$$\Sigma v_R \frac{\partial}{\partial R}(Rv_\theta) = \frac{1}{R}\frac{\partial}{\partial R}\left(R^3 \Sigma v \frac{\partial}{\partial R}\frac{v_\theta}{R}\right). \qquad (3.5.13)$$

Also, in taking v_θ to be time-independent we have already had to assume that the disk is much less massive than the central mass. The orbits are therefore

Keplerian, with $v_\theta \propto R^{-1/2}$, so on the left-hand side of Eq. (3.5.13) we can use $(\partial/\partial R)(Rv_\theta) = v_\theta/2$, and on the right-hand side use

$$\frac{\partial}{\partial R}\left(R^3 \Sigma v \frac{\partial}{\partial R}\frac{v_\theta}{R}\right) = -\frac{3}{2}\frac{\partial}{\partial R}(R\Sigma v v_\theta) = -\frac{3}{2}R^{1/2}v_\theta \frac{\partial}{\partial R}\left(R^{1/2}\Sigma v\right).$$

Canceling factors of v_θ in Eq. (3.5.13), we have then

$$\Sigma v_R = -3R^{-1/2}\frac{\partial}{\partial R}\left(R^{1/2}\Sigma v\right). \tag{3.5.14}$$

We can use this to eliminate v_R in Eq. (3.5.11), and have at last our differential equation for the surface density:

$$\frac{\partial \Sigma}{\partial t} = \frac{3}{R}\frac{\partial}{\partial R}\left(\sqrt{R}\frac{\partial}{\partial R}(\sqrt{R}\Sigma v)\right). \tag{3.5.15}$$

We will explore some time-dependent solutions of Eq. (3.5.15) at the end of this section, making a simple assumption about the viscosity v. For the present, we will use Eq. (3.5.15) to derive formulas for the flow of mass and angular momentum through the disk, that hold for arbitrary viscosity. We will then use these formulas in a treatment of the most interesting special case, a steady disk with Σ time-independent.

To calculate the rates of flow of mass and angular momentum, we consider the rate of change of the amount of mass and angular momentum of the disk between any two radii, R_1 and R_2. Using Eq. (3.5.15), we find the rate of change of the mass $M(R_1, R_2)$ between radii R_1 and R_2:

$$\begin{aligned}\frac{dM(R_1,R_2)}{dt} &= \frac{d}{dt}\int_{R_1}^{R_2} 2\pi R\, dR\, \Sigma(R,t) \\ &= 6\pi\left[R^{1/2}\frac{\partial}{\partial R}(R^{1/2}\Sigma v)\right]_{R=R_2} \\ &\quad - 6\pi\left[R^{1/2}\frac{\partial}{\partial R}(R^{1/2}\Sigma v)\right]_{R=R_1}.\end{aligned} \tag{3.5.16}$$

This tells us that the rate of mass flow inward through a circle of radius R is

$$\dot{M}(R,t) = 6\pi R^{1/2}\frac{\partial}{\partial R}(R^{1/2}\Sigma v). \tag{3.5.17}$$

This result is pretty obvious, for using Eq. (3.5.14) allows us to rewrite it as

$$\dot{M}(R,t) = -2\pi R\Sigma(R,t)v_R(R,t).$$

Naturally, for v_R negative, in a time dt particles at R move inward a distance $-v_R\, dt$, and the mass of the particles in an annulus of this thickness is $-v_R\, dt \times 2\pi R\Sigma$.

The result for angular momentum flow is less obvious. As we have noted, $v_\theta(R)$ is time-independent at a fixed R, so the θ-component (3.5.10) of the equations of motion gives

$$R^2 v_\theta \frac{\partial \Sigma}{\partial t} = \frac{\partial}{\partial R}\left[-R^2 \Sigma v_R v_\theta + R^3 \Sigma \nu \frac{\partial}{\partial R}\frac{v_\theta}{R}\right],$$

and the rate of change of the angular momentum $J(R_1, R_2)$ between R_1 and R_2 is therefore

$$\frac{dJ(R_1,R_2)}{dt} = \frac{d}{dt}\int_{R_1}^{R_2} 2\pi R\, dR \times \Sigma(R,t) R v_\theta(R)$$

$$= 2\pi \left[-R^2 \Sigma v_R v_\theta + R^3 \Sigma \nu \frac{\partial}{\partial R}\frac{v_\theta}{R}\right]_{R=R_2}$$

$$- 2\pi \left[-R^2 \Sigma v_R v_\theta + R^3 \Sigma \nu \frac{\partial}{\partial R}\frac{v_\theta}{R}\right]_{R=R_1}. \tag{3.5.18}$$

We conclude that the rate at which angular momentum moves inward through a circle of any radius R is

$$\dot{J}(R,t) = 2\pi\left[-R^2 \Sigma v_R v_\theta + R^3 \Sigma \nu \frac{\partial}{\partial R}\frac{v_\theta}{R}\right].$$

Using Eq. (3.5.14) and $v_\theta \propto R^{-1/2}$, this is

$$\dot{J}(R,t) = 2\pi v_\theta\left[3R^{3/2}\frac{\partial}{\partial R}(R^{1/2}\Sigma\nu) - \frac{3}{2}R\Sigma\nu\right]$$

$$= 6\pi v_\theta(R) R^2 \frac{\partial}{\partial R}\bigl(\Sigma(R,t)\nu(R,t)\bigr). \tag{3.5.19}$$

Steady Disks

We now consider the case of an accretion disk that has settled into a quasi-stable configuration that lasts much longer than the characteristic time R^2/ν. Since matter and angular momentum are assumed to be accreting onto a central body, in this case we must assume that they are being supplied by matter coming in to the disk from outside it. For a steady disk we can drop the time derivative in Eq. (3.5.15). The solution of Eq. (3.5.15) for $\Sigma\nu$ then is trivial:

$$\Sigma\nu = \frac{A}{\sqrt{R}} + B, \tag{3.5.20}$$

where A and B are constants.

The constant B can be related to the rate $\dot{M} \equiv -2\pi R\Sigma v_R$ at which mass flows inward through the $2\pi R$ circumference of a circle of any radius R. Inserting Eq. (3.5.20) in Eq. (3.5.17) gives

$$\dot{M} = 3\pi B. \tag{3.5.21}$$

Likewise, inserting Eq. (3.5.20) into Eq. (3.5.19) gives the rate of inward angular momentum flow

$$\dot{J} = -3\pi A v_\theta R^{1/2} = -3\pi A \sqrt{MG}. \qquad (3.5.22)$$

Note that for steady disks both $\dot{M}$ and $\dot{J}$ are independent of both t and R, as of course they must be if the disk properties are not to change. These constants depend on the environment of the disk – the amount of matter and angular momentum flowing in from the outer edge of the disk.

We will now make the additional assumption that the central body is not rotating, or in any case not rotating rapidly enough to exert an appreciable torque on the disk. (That is, the velocity of rotation at its surface must be much less than the velocity a satellite would have in a low orbit, as is the case for the Earth and the Sun.) This allows us to decide about a crucial sign difference between A and B. For a central body that is not a source of mass or angular momentum, both $\dot{M}$ and $\dot{J}$ (which are defined as rates of inward flow) must have the same sign as M and J, and hence be positive. It follows then from (3.5.21) and (3.5.22) that while B is positive, A is *negative*.[27] With B positive and A negative, there is a radius R_0 at which $\Sigma \nu$ vanishes:

$$+ R_0^{1/2} \equiv -A/B, \qquad (3.5.23)$$

and we can write Eq. (3.5.20) as

$$\Sigma \nu = \frac{\dot{M}}{3\pi} \left(1 - (R_0/R)^{1/2} \right). \qquad (3.5.24)$$

For an accretion disk around a star, the orbit of radius R_0 may be taken as just above the star's surface. On the other hand, as discussed in Section 4.5, for an accretion disk around a black hole R_0 is instead taken as the radius of the smallest stable orbit around the black hole.

It may be noted that since $\Sigma \nu$ vanishes at $R = R_0$, the angular momentum inflow $\dot{J}$ is given by the first term in the first expression in Eq. (3.5.19) as

$$\dot{J} = 6\pi R_0^{3/2} v_\theta(R_0) \left[\frac{\partial}{\partial R} \left(R^{1/2} \Sigma \nu \right) \right]_{R=R_0},$$

which according to Eq. (3.5.17) is just $R_0 v_\theta(R_0) \dot{M}$. Hence the ratio of $\dot{J}$ to $\dot{M}$ equals the ratio of angular momentum to mass of the particles falling onto the central body from their orbit at R_0.

Nevertheless, it is *not* correct to suppose that the inward angular momentum flow $\dot{J}$ is simply due to the transport inwards of the particles that carry the mass flow $\dot{M}$. The angular momentum of a particle of mass m in an orbit of radius

---

[27] Note that in deriving Eq. (3.5.19) we assumed that the angular momentum in an annulus of thickness dR is $2\pi R \, dR \, \Sigma R v_\theta$, so our convention for J is that it is positive if v_θ is, and therefore by taking v_θ positive we should get a positive $\dot{J}$ in Eq. (3.5.22).

R is $j = mRv_\theta(R)$, so $j/m = Rv_\theta(R) = \sqrt{RMG}$. For any radius $R \gg R_0$ this is much greater than $\dot{J}/\dot{M}$, which according to Eqs. (3.5.21)–(3.5.23) is $\sqrt{R_0 MG}$. So what happens to the angular momentum of the particles flowing in to the center of the disk? Recall that these particles flow inward because of the effects of viscosity on a differentially rotating disk. The particles within a circle of any radius R are moving faster than the particles outside the circle, so viscosity produces a torque that reduces the speed of the particles inside the circle, which then move inward. At the same time, this torque tends to increase the speed of the particles outside the circle, so that angular momentum flows outward. In an isolated disk with no source of angular momentum flowing in at the outer edge the net flow of angular momentum would have to be outward, but such a disk could not be time-independent, because as already mentioned there is no source of angular momentum at the center. In a steady disk whose edge radius is much larger than R_0, whatever angular momentum flows in from the edge is nearly canceled by the angular momentum flowing outward because of viscous torque, leaving $\dot{J}/\dot{M}$ much less than the ratio of angular momentum to mass of individual particles.

Viscosity causes not only the transport of mass and angular momentum in accretion disks – it also heats the disks. Fortunately, as we shall see, this can be calculated knowing only $\Sigma \nu$, without separate information about Σ or ν. The general formula for the rate per volume of heat produced by viscosity in imperfect fluids is[28]

$$\mathcal{H} = \frac{\rho \nu}{2} \sigma_{ij} \sigma_{ij} + \zeta (\nabla \cdot \mathbf{v})^2, \tag{3.5.25}$$

where σ_{ij} is the tensor multiplying $\rho \nu$ in Eq. (3.5.6):

$$\sigma_{ij} = \frac{\partial v_i}{\partial x_j} + \frac{\partial v_j}{\partial x_i} - \frac{2}{3}\delta_{ij}\left(\frac{\partial v_k}{\partial x_k}\right).$$

In cylindrical coordinates, taking into account only the θ-component of velocity, this gives

$$\mathcal{H} = \rho \nu \left(\frac{\partial v_\theta}{\partial R} - \frac{v_\theta}{R}\right)^2 = \rho \nu R^2 (\Omega'(R))^2, \tag{3.5.26}$$

and the corresponding heat production per disk area is

$$\int dz\, \mathcal{H} = \Sigma \nu R^2 (\Omega'(R))^2. \tag{3.5.27}$$

We can find an explicit formula for the R-dependence of the rate of heat production per area by using Eq. (3.5.24) for $\Sigma \nu$ and the Newton formula $\Omega^2 = GM/R^3$ (where M is the central mass) for Ω':

---

[28] L. D. Landau and E. M. Lifshitz, *Fluid Mechanics*, trans. J B. Sykes and W. H. Reid (Pergamon Press, London, 1959), Eq. (49.5).

$$\int dz\, \mathcal{H} = \frac{9GM}{4R^3} \frac{\dot{M}}{3\pi} \left(1 - (R_0/R)^{1/2}\right). \tag{3.5.28}$$

The total rate of heat production over the whole disk, from radius R_0 to a much larger radius, is

$$\int_{R_0}^{\infty} 2\pi R\, dR \int dz\, \mathcal{H} = \frac{GM\dot{M}}{2R_0}. \tag{3.5.29}$$

A mass m that falls from rest at infinity to rest at R_0 loses an energy GMm/R_0 to gravitation, so Eq. (3.5.29) tells us that half of that energy goes into viscous heating of the disk.

This formalism allows us to calculate not only the total luminosity produced by viscosity, but also its frequency distribution. Just as we did with stars, if we assume that all heat is radiated we can define an effective radiation temperature $T_{\text{eff}}(R)$, by setting the rate (3.5.28) of heat production per area equal to the rate σT_{eff}^4 per area at which a black body will radiate energy, so that

$$T_{\text{eff}}(R) = \left(\frac{3GM\dot{M}}{4\pi\sigma R^3}\left[1 - (R_0/R)^{1/2}\right]\right)^{1/4}, \tag{3.5.30}$$

where $\sigma = 2\pi^5 k^4/15h^3c^3$ is the Stefan–Boltzmann constant. Assuming that the disk is optically thick, the rate per area at which energy is radiated by the disk at R between frequencies ν and $\nu + d\nu$ is then $2B(\nu, T_{\text{eff}}(R))$, where $B(\nu, T)$ is given by the Planck formula

$$B(\nu, T)\, d\nu = \frac{2\pi h}{c^2} \frac{\nu^3\, d\nu}{\exp(h\nu/k_B T) - 1}, \tag{3.5.31}$$

and the factor 2 appears multiplying B because the disk has two sides. Where we can't resolve the disk, as for a quasi-stellar source or object, it is only the absolute luminosity $L(\nu)\, d\nu$ between frequencies ν and $\nu + d\nu$ that is available for observation. It is given by an integral over the area of the disk

$$L(\nu) = \int 2\pi R\, dR\, B(\nu, T_{\text{eff}}(R)). \tag{3.5.32}$$

(The factor 2 is omitted here because we see only one side of the disk.) Although this differs from a simple black-body spectrum, with a single temperature, it is not generally easy to tell the difference. The best evidence in support of the spectrum given by Eqs. (3.5.30)–(3.5.32) currently comes from the study of cataclysmic variables, described at the end of this section.

A complication must be mentioned. The heat generated by viscosity in the outer layers of a disk may not be enough to raise that part of the disk to a sufficient temperature to allow this heat to be efficiently radiated away before it is carried inward by the radial motion of the disk, like an impurity being advected by an eddy in a river. Where this happens, the observed temperature in the inner part of the disk will be higher than given by Eq. (3.5.30), with an observable effect on the spectrum of the whole disk.

Before going on to consider time-dependent solutions, we will take a moment to consider the vertical structure of the disk. It is set by a balance between the z-components of gravitational and pressure forces:

$$\frac{1}{\rho}\frac{\partial p}{\partial z} = -\frac{MGz}{(R^2+z^2)^{3/2}} \simeq -\frac{MGz}{R^3}. \tag{3.5.33}$$

The speed of sound c_s is given by $c_s^2 = \partial p/\partial \rho$, so Eq. (3.5.33) reads

$$\frac{1}{\rho}\frac{\partial \rho}{\partial z} = -\frac{MGz}{c_s^2 R^3}. \tag{3.5.34}$$

(The meaning of c_s depends on circumstances. For an ideal gas with fixed temperature p is proportional to ρ, and so $c_s^2 = p/\rho$. For adiabatic variations in a polytrope of index Γ we have $p \propto \rho^\Gamma$, so $c_s^2 = \Gamma p/\rho$. Considering the approximations made here, a factor Γ is not very important.) Taking the sound speed as a constant, the solution is a Gaussian

$$\rho \propto \exp\left(-\left(\frac{MG}{2c_s^2 R^3}\right)z^2\right). \tag{3.5.35}$$

(Actually, just as for a star, the disk's temperature decreases toward its surface, so c_s decreases toward the surface, and the fall-off of density is even faster than for a Gaussian.) From Eq. (3.5.35) we see that the thickness of the disk is of order

$$\Delta z \approx \sqrt{\frac{c_s^2 R^3}{MG}} = \frac{c_s}{\Omega}, \tag{3.5.36}$$

where as usual $\Omega = (MG/R^3)^{1/2}$ is the orbital frequency. Thus the thin-disk approximation $\Delta z \ll R$ requires $c_s \ll \Omega R$. In other words, the orbital motion must be *supersonic*.

This is part of the modern standard picture of accretion disks, due largely to Shakura and Sunyaev.[29] They and their successors go into detail regarding the internal structure of the disk and its stability, too much detail for us here.

Decaying Disks

We now return to the more general problem, of time-dependent accretion disks. We now assume that the disk extends to a great distance, where the surface density Σ becomes negligible, with nothing coming in from infinity. Since mass and angular momentum are falling onto the central body, the density of the disk at any given radius R will decay. But as time passes the matter of the

---

[29] N. I. Shakura and R. A. Sunyaev, *Astron. Astrophys.* **24**, 357 (1973).

disk will extend to greater distances. Indeed, by setting $R_1 = 0$ and $R_2 = \infty$ in Eq. (3.5.18), we see that the total angular momentum of the disk remains constant.

In this case, we cannot get much use from Eq. (3.5.15) without knowing something about the viscosity ν. In general ν depends on density and temperature, and to find these we need to solve equations of energy as well as mass and momentum transport. If, as seems plausible, the viscosity depends on Σ, Eq. (3.5.15) is not even linear.

If we somehow could find ν as a function only of R, then it would be possible to make progress in finding solutions of Eq. (3.5.15) by looking for a factorized solution. Suppose we try

$$\Sigma(R,t) = f(R)g(t).$$

Dividing Eq. (3.5.15) by fg, we have

$$\frac{\dot{g}(t)}{g(t)} = \frac{3}{Rf(R)} \frac{d}{dR}\left(\sqrt{R}\frac{d}{dR}\left(\sqrt{R}\nu(R)f(R)\right)\right).$$

One side of this equation is independent of R, while the other is independent of t, so both must just be a constant, say $-\omega$. Setting the left-hand side equal to $-\omega$ gives $g(t) \propto \exp(-\omega t)$, while setting the right-hand side equal to $-\omega$ gives

$$\frac{d}{dR}\left(\sqrt{R}\frac{d}{dR}\left(\sqrt{R}\nu(R)f(R)\right)\right) = \frac{-\omega R f(R)}{3}.$$

For instance, if we make the simple though quite unrealistic assumption that ν is constant, then with $\omega > 0$ the two solutions for $f(R)$ are $R^{-1/4}J_{\pm 1/4}(\sqrt{\omega/3\nu}R)$, one of which,

$$f(R) \propto R^{-1/4}J_{1/4}(\sqrt{\omega/3\nu}R),$$

is non-singular at $R = 0$. The general non-singular decaying solution is a superposition, which (introducing $u \equiv \sqrt{\omega/3\nu}$) may be written

$$\Sigma(R,t) = R^{-1/4}\int_0^\infty du\, g(u) \exp\left(-3\nu u^2 t\right) J_{1/4}(uR). \qquad (3.5.37)$$

An 1875 result[30] of Hermann Hankel lets us evaluate $g(u)$ in terms of the surface density at $t = 0$:

$$g(u) = u \int_0^\infty dr\, r^{5/4} \Sigma(r,0) J_{1/4}(ur). \qquad (3.5.38)$$

We can use this to give a simple formula for the late-time behavior of the surface density. We note first that for $t \to \infty$ the integral (3.5.37) is dominated by

---

[30] G. N. Watson, *A Treatise on the Theory of Bessel Functions*, 2nd edn. (Cambridge University Press, Cambridge, 1944), p. 453.

values of u for which u is no greater than of order $1/\sqrt{3vt}$. For $R/\sqrt{3vt} \ll 1$ we can use the limiting formula $J_{1/4}(z) \to (z/2)^{1/4}/\Gamma(5/4)$ for $z \to 0$, and find

$$\Sigma(R,t) \to \frac{1}{\Gamma(5/4)} \int_0^\infty du\; g(u)\, (u/2)^{1/4} \exp\left(-3vu^2 t\right). \qquad (3.5.39)$$

Since this integral is dominated by values of u for which u is not much greater than $1/\sqrt{3vt}$, we can use the limit of Eq. (3.5.38) for small u:

$$g(u) \to \frac{u^{3/4}}{2^{1/4}\Gamma(5/4)} \int_0^\infty dr\; r^{3/2} \Sigma(r,0). \qquad (3.5.40)$$

The remaining integral is proportional to the constant total angular momentum of the disk:

$$J = \int_0^\infty 2\pi r\, dr \times r v_\theta(r) \Sigma(r,0) = 2\pi \sqrt{MG} \int_0^\infty dr\; r^{3/2} \Sigma(r,0). \qquad (3.5.41)$$

Using Eqs. (3.5.39)–(3.5.41), we have at late times, for $R \ll \sqrt{3vt}$,

$$\Sigma(R,t) \to \frac{1}{2\sqrt{2}\Gamma(5/4)(3vt)^{5/4}} \int_0^\infty dr\; r^{3/2} \Sigma(r,0)$$

$$= \frac{J}{4\pi\sqrt{2}\Gamma(5/4)\sqrt{MG}(3vt)^{5/4}}. \qquad (3.5.42)$$

(The same result may be obtained more directly from a less obvious solution[31] of Eq. (3.5.15):

$$\Sigma(R,t) = \frac{2}{R} \int_0^\infty dr\; \Sigma(r,0) \frac{x^{3/4}}{\tau} \exp\left(-\frac{1+x^2}{\tau}\right) I_{1/4}\left(\frac{2x}{\tau}\right), \qquad (3.5.43)$$

where $x \equiv R/r$, $\tau \equiv 12vt/r^2$, and $I_{1/4}(z)$ is the modified Bessel function, $I_\nu(z) \equiv e^{-\nu\pi i} J_\nu(z e^{\pi i/2})$ for $-\pi < \mathrm{Arg}\, z < \pi/2$.)

The R-independence and t-dependence of the limiting formula (3.5.42) is not in conflict with the finiteness and time-independence of the angular momentum J, because this formula is only supposed to hold for $R \ll \sqrt{3vt}$. The disk does not have a sharp edge, but we may define an effective radius $\mathcal{R}(t)$ such that the angular momentum is the same as if $\Sigma(R,t)$ were correctly given by Eq. (3.5.42) for $R < \mathcal{R}(t)$ and were zero for $R > \mathcal{R}(t)$. That is, $\mathcal{R}(t)$ is defined so that

$$\int_0^{\mathcal{R}(t)} R^{3/2} \left[\frac{1}{2\sqrt{2}\Gamma(5/4)(3vt)^{5/4}} \int_0^\infty dr\; r^{3/2} \Sigma(r,0) \right] dR$$

$$= \int_0^\infty dR\; R^{3/2} \Sigma(R,0). \qquad (3.5.44)$$

---

[31] J. E. Pringle, *Ann. Rev. Astron. Astrophys.* **19**. 137 (1981).

This gives our result for the growth of the disk's effective radius:

$$\mathcal{R}(t) \to \left(5\sqrt{2}\Gamma(5/4)\right)^{2/5} \sqrt{3\nu t}. \qquad (3.5.45)$$

The detailed formulas (3.5.42) and (3.5.45) for the decay and spread of the disk depend on the assumption of constant viscosity, which as already mentioned is not realistic. But these formulas serve to illustrate the point that an accretion disk that is not taking in matter from outside must decay, and since its angular momentum is constant, it must also spread.

Accretion Disks in Binaries

Accretion disks are formed not only in the accretion of matter from the interstellar medium, but also when matter is accreted on a compact object from a companion star in a binary that has filled its Roche lobe. Among these are the cataclysmic variables, binaries consisting of a white dwarf star that is accreting matter from an ordinary star. (The "cataclysms" occur sporadically when nuclear energy is released from the conversion into helium of hydrogen that has accreted onto the white dwarf, an event often called a nova.) As mentioned earlier, these cataclysmic variables provide the best evidence for the spectrum shape expected for radiation emitted from steady accretion disks. The compact object may instead be a black hole. A famous example is the X-ray source Cygnus X-1, discovered in 1964. In Cygnus X-1 matter is flowing from a supergiant star of type O7 onto a black hole with a mass about $10 M_\odot$. By now dozens of such black holes in binaries have been discovered. Because white dwarfs and black holes are so compact, the general theory of accretion disks described in this section can be applied to accretion disks in binaries, but observations are complicated by the presence of the companion star.

3.6 Accretion Spheres

The accretion disks discussed in Section 3.5 arise when angular momentum as well as mass is accreted by a central body from the interstellar medium. Accretion is also possible without forming a disk, when the interstellar gas surrounding a central body is pretty much at rest, carrying little angular momentum along with its mass. As an idealization, we can treat this sort of accretion as a steady spherically symmetric infall of gas from interstellar matter that at large distances from the central body is at rest, with position-independent density and pressure. Instead of the attraction of gravitation being resisted by centrifugal force, as in accretion disks, here it is resisted by gas pressure. In the absence of differential rotation, viscosity can be neglected. This sort of spherically symmetric inviscid accretion was first analyzed by Herman Bondi in

1952,[32] before the modern work on accretion disks described in Section 3.5. It is believed to account for accretion onto various neutron stars and black holes, such as the stellar-mass black hole M31*, which is observed as an X-ray and radio source in the Andromeda galaxy.

We start with the equation of motion (3.5.6), which in the absence of viscosity reads

$$\frac{\partial}{\partial t}(\rho v_i) + \frac{\partial}{\partial x_j}(\rho v_i v_j) = -\rho \frac{\partial \phi}{\partial x_i} - \frac{\partial p}{\partial x_i}, \quad (3.6.1)$$

and the equation of continuity (3.5.7)

$$\frac{\partial \rho}{\partial t} + \frac{\partial}{\partial x_i}(\rho v_i) = 0. \quad (3.6.2)$$

As before, ρ is the gas mass density; v_i are the Cartesian components of the gas velocity; ϕ is the gravitational potential; p is the gas pressure; i and j run over the Cartesian coordinate directions; and repeated indices are summed. We use Eq. (3.6.2) to put Eq. (3.6.1) in a more convenient form:

$$\frac{\partial v_i}{\partial t} + v_j \frac{\partial v_i}{\partial x_j} = -\frac{\partial \phi}{\partial x_i} - \frac{1}{\rho}\frac{\partial p}{\partial x_i}. \quad (3.6.3)$$

We specialize here to the case of steady flow and spherical symmetry, in which the velocity has only a radial component $v \equiv v_r$, and v, p, ρ, and ϕ depend only on the radial coordinate r. The equation of continuity (3.6.2) then takes the form

$$\frac{1}{r^2}\frac{d}{dr}(r^2 \rho v) = 0, \quad (3.6.4)$$

and the only non-trivial component of the equation of motion (3.6.3) is the r-component

$$v \frac{dv}{dr} = -\frac{d\phi}{dr} - \frac{1}{\rho}\frac{dp}{dr}. \quad (3.6.5)$$

Equation (3.6.4) lets us write

$$4\pi r^2 \rho v = -\dot{M}, \quad (3.6.6)$$

where $\dot{M}$ is the r-independent accretion rate. (The minus sign is inserted here because in accretion v is negative, so that $\dot{M}$ is positive.) The equation of motion can also be written as a conservation law, but first we have to consider the dependence of pressure on density. If we suppose that heat conduction as well as viscosity is negligible then variations in pressure and density are adiabatic, and, as discussed in Section 1.8, for a variety of gases they are related by a polytropic equation of state:

$$p = K\rho^\Gamma, \quad (3.6.7)$$

---

[32] H. Bondi, *Mon. Not. Roy. Astron. Soc.* **112**, 195 (1952).

with K and Γ constants. Then

$$\frac{1}{\rho}\frac{dp}{dr} = \frac{K\Gamma}{\Gamma-1}\frac{d\rho^{\Gamma-1}}{dr}, \qquad (3.6.8)$$

so Eq. (3.6.5) tells us that

$$\frac{d}{dr}\left(\frac{v^2(r)}{2} + \phi(r) + \frac{K\Gamma\rho^{\Gamma-1}(r)}{\Gamma-1}\right) = 0. \qquad (3.6.9)$$

We take the gravitational potential for a central body of mass M as $\phi(r) = -MG/r$. At an infinite distance from the central body $\rho(r)$ approaches some constant value $\rho(\infty)$, while $v(r)$ and $\phi(r)$ vanish, so Eq. (3.6.9) can be expressed as

$$\frac{v^2(r)}{2} - \frac{MG}{r} + \frac{K\Gamma\rho^{\Gamma-1}(r)}{\Gamma-1} = \frac{K\Gamma\rho^{\Gamma-1}(\infty)}{\Gamma-1}. \qquad (3.6.10)$$

We will now assume, as is typically the case, that the accretion is *transonic*. That is, while $v(r)$ is much less than the ambient sound speed $c_s(r)$ at very large distances r, the gas speed $|v(r)|$ rises to a supersonic value much greater than $c_s(r)$ as $r \to 0$, so there must be a critical radius r_c at which

$$v(r_c) = -c_s(r_c). \qquad (3.6.11)$$

The pair of algebraic equations (3.6.10) and (3.6.6) can be solved to find $v(r)$ and $\rho(r)$ as functions of r for a broad range of values of the accretion rate $\dot{M}$, but none of these solutions correspond to a stable transonic accretion, except the one for a particular critical value of $\dot{M}$. Accretion is possible for any value of $\dot{M}$, but not *stable transonic* accretion. For any value of $\dot{M}$ other than the critical value, transonic accretion is time-dependent, and can become stable only if $\dot{M}$ evolves to its critical value.

In general under adiabatic conditions for a polytrope with index Γ the speed of sound is

$$c_s(r) = \left(\frac{\partial p}{\partial \rho}\right)_{\text{ad}}^{1/2} = \left(K\Gamma\rho^{\Gamma-1}(r)\right)^{1/2} = c_s(\infty)\left(\frac{\rho(r)}{\rho(\infty)}\right)^{(\Gamma-1)/2}. \qquad (3.6.12)$$

To see what the transonic condition implies for $\dot{M}$, let us return to the equation of motion (3.6.5), this time using Eqs. (3.6.6) and (3.6.7) to write

$$\frac{1}{\rho}\frac{dp}{dr} = \frac{c_s^2}{\rho}\frac{d\rho}{dr} = -\frac{c_s^2}{r^2 v}\frac{d(r^2 v)}{dr} = -\frac{c_s^2}{v}\frac{dv}{dr} - \frac{2c_s^2}{r}, \qquad (3.6.13)$$

so (3.6.5) becomes

$$\left(1 - \frac{c_s^2(r)}{v^2(r)}\right)v(r)\frac{dv(r)}{dr} + \frac{MG}{r^2} - \frac{2c_s^2(r)}{r} = 0. \qquad (3.6.14)$$

This is known as the *wind equation*, because it was first encountered in the seminal study of stellar winds by Eugene Parker.[33]

For accretion or stellar winds, we have respectively $v(r) = -c_s(r)$ or $v(r) = +c_s(r)$ at the critical radius $r = r_c$, which according to Eq. (3.6.14) satisfies

$$\frac{MG}{2r_c c_s^2(r_c)} = 1. \tag{3.6.15}$$

As we have seen, for accretion the boundary conditions at infinite r give Eq. (3.6.10), which with Eq. (3.6.12) may be written

$$\frac{v^2(r)}{2} - \frac{MG}{r} + \frac{c_s^2(r)}{\Gamma - 1} = \frac{c_s^2(\infty)}{\Gamma - 1}. \tag{3.6.16}$$

Setting $r = r_c$ then gives

$$c_s^2(r_c) \left(\frac{1}{2} - 2 + \frac{1}{\Gamma - 1} \right) = \frac{c_s^2(\infty)}{\Gamma - 1},$$

or in other words

$$c_s^2(r_c) = \left(\frac{2}{5 - 3\Gamma} \right) c_s^2(\infty). \tag{3.6.17}$$

We can now calculate the accretion rate in terms of quantities evaluated far from the accreting body, by setting $r = r_c$ in Eq. (3.6.6). Using Eq. (3.6.12) and the defining conditions $v(r_c) = -c_s(r_c)$ and (3.6.15) in Eq. (3.6.6) gives

$$\dot{M} = 4\pi r_c^2 \rho(r_c) c_s(r_c) = 4\pi \left(\frac{MG}{2c_s^2(r_c)} \right)^2 \rho(\infty) \left(\frac{c_s(r_c)}{c_s(\infty)} \right)^{2/(\Gamma-1)} c_s(r_c)$$

$$= \pi M^2 G^2 \rho(\infty) c_s^{-3}(\infty) \left(\frac{c_s(r_c)}{c_s(\infty)} \right)^{(5-3\Gamma)/(\Gamma-1)}.$$

Using Eq. (3.6.17), we now find

$$\dot{M} = \pi M^2 G^2 \rho(\infty) c_s^{-3}(\infty) \left(\frac{5 - 3\Gamma}{2} \right)^{-(5-3\Gamma)/2(\Gamma-1)}. \tag{3.6.18}$$

Thus, knowing the ambient density and sound speed, and the value of the polytrope index Γ, we can find the unique rate of stable Bondi accretion. For a monatomic ideal gas with $\Gamma = 5/3$ the final factor in Eq. (3.6.18) is $0^0 = 1$. In terms of reference values for $\rho(\infty)$ and $c_s(\infty)$ typical of the environment of M31*,[34] $\rho(\infty) \approx 10^{-24}$ g/cm^3 and $c_s(\infty) \approx 10$ km/sec, the accretion rate is

$$\dot{M} \approx 5.5 \times 10^{10} \left(\frac{M}{M_\odot} \right)^2 \left(\frac{\rho(\infty)}{10^{-24} \text{ g/cm}^3} \right) \left(\frac{c_s(\infty)}{10 \text{ km/sec}} \right)^{-3} \text{ g/sec}.$$

---

[33] E. Parker, *Astrophys. J.* **132**, 821 (1960).
[34] These numbers are taken from F. Melia, *High-Energy Astrophysics* (Princeton University Press, Princeton, NJ, 2009).

If a fraction f of the accreted mass is converted to energy, the luminosity is

$$f\dot{M}c^2 = 5 \times 10^{31} f \left(\frac{M}{M_\odot}\right)^2 \left(\frac{\rho(\infty)}{10^{-24}\,\mathrm{g/cm^3}}\right) \left(\frac{c_s(\infty)}{10\,\mathrm{km/sec}}\right)^{-3} \mathrm{erg/sec}.$$

The observed X-ray luminosity of M31* is about 3×10^{35} erg/sec, suggesting a mass equal to a few tens of solar masses.

3.7 Soft Bremsstrahlung

In this section we consider the emission of observable radiation due to bremsstrahlung, produced when a free electron is scattered by an atomic nucleus in an atom or ion. The consideration of observable bremsstrahlung radiation will require a treatment different from our earlier discussion of related processes: In calculating the Rosseland mean free–free opacity in Section 1.4 we averaged over the frequencies of absorbed photons, and in calculating bremsstrahlung cooling in Section 3.3 we integrated over the frequencies of emitted photons. The results in both calculations were largely dominated by unobserved photons with energies $h\nu$ roughly of order $k_B T$. Here instead we want to calculate the observable emissivity $j_\nu(T)$, the energy emitted at temperature T per time, per volume, per solid angle, and per frequency interval, at a specific frequency ν. We will be chiefly interested in soft bremsstrahlung, for which $h\nu \ll k_B T$, as is typically the case for the radio waves from hot ionized gas in galaxies and galaxy clusters.

The emissivity is calculated in terms of the rate $j(\nu, v)$ of emission per electron of velocity $\mathbf{v}$, per second, per photon solid angle, and per photon frequency interval:

$$j_\nu(T) = \int_{m_e v^2/2 > h\nu} d^3v \, n_e(\mathbf{v}, T) j(\nu, |\mathbf{v}|), \tag{3.7.1}$$

where $n_e(\mathbf{v}, T) \, d^3v$ is the number density of free electrons with velocity $\mathbf{v}$ in a range d^3v at $\mathbf{v}$. It is conventional to express $j(\nu, v)$ as the approximate classical electrodynamics result given in 1923 by Kramers,[35] times a "free–free Gaunt factor" $g_{\mathrm{ff}}(\nu, v)$ that incorporates quantum and other corrections:

$$j(\nu, v) \equiv \frac{8\pi Z^2 e^6 n_N}{3\sqrt{3}\, c^3 m_e^2 v} g_{\mathrm{ff}}(\nu, v), \tag{3.7.2}$$

where n_N is the number density of ions, Ze is the ionic charge (with e everywhere in unrationalized electrostatic units), and m_e is the electron mass. Using

---
[35] H. Kramers, *Phil. Mag.* **46**, 836 (1923).

3.7 Soft Bremsstrahlung

the Maxwell–Boltzmann momentum-space distribution (3.3.24) for $n_e(\mathbf{v}, T)$, this gives the emissivity

$$j_\nu(T) = \frac{8Z^2 e^6 n_N n_e}{3c^3(k_B T)^{1/2} m_e^{3/2}} \left(\frac{2\pi}{3}\right)^{1/2} \overline{g}_{\text{ff}}(\nu, T), \quad (3.7.3)$$

where n_e is the total number density of free electrons; k_B is the Boltzmann constant; and $\overline{g}_{\text{ff}}(\nu, T)$ is the thermally averaged free–free Gaunt factor (or, briefly, the thermal Gaunt factor):

$$\overline{g}_{\text{ff}}(\nu, T) = \frac{m_e}{k_B T} \int_{\sqrt{2h\nu/m_e}}^\infty g_{\text{ff}}(\nu, v) \exp\left(-\frac{m_e v^2}{2k_B T}\right) v\, dv. \quad (3.7.4)$$

Astrophysicists today chiefly rely on various numerical calculations[36] of the Gaunt factor, based on a set of quite complicated formulas:

$$g_{\text{ff}}(\nu, v) = \frac{2\sqrt{3}}{\pi \xi \xi'} \left[(\xi^2 + \xi'^2 + 2\xi^2 \xi'^2) I_0 - 2\xi\xi'(1+\xi^2)^{1/2}(1+\xi'^2)^{1/2} I_1 \right] I_0, \quad (3.7.5)$$

where

$$I_\ell = \frac{1}{4} \left(\frac{4\xi\xi'}{(\xi' - \xi)^2}\right)^{\ell+1} e^{\pi(\xi+\xi')/2} \frac{|\Gamma(\ell + 1 + i\xi)\Gamma(\ell + 1 + i\xi')|}{\Gamma(2\ell + 1)}$$

$$\times \left(\frac{\xi + \xi'}{\xi' - \xi}\right)^{-i\xi - i\xi'} {}_2F_1\left(\ell + 1 - i\xi, \ell + 1 - i\xi'; 2\ell + 2; -\frac{4\xi\xi'}{(\xi' - \xi)^2}\right). \quad (3.7.6)$$

Here $\xi \equiv Ze^2/\hbar v$ and $\xi' \equiv Ze^2/\hbar v'$, with v' the magnitude of the final electron velocity, given in terms of ν and v by the condition of energy conservation

$$m_e v'^2/2 = m_e v^2/2 - h\nu. \quad (3.7.7)$$

Also, ${}_2F_1$ is a confluent hypergeometric function, with power series expansion

$${}_2F_1(a, b; c; x) = \sum_{n=0}^\infty \frac{(a)_n (b)_n}{(c)_n} \frac{x^n}{n!}, \quad (3.7.8)$$

where for any complex z

$$(z)_n \equiv z(z+1)\cdots(z+n-1) \text{ for } n = 1, 2, 3, \ldots; \quad (z)_0 \equiv 1.$$

---

[36] For instance, W. J. Karzas and R. Latter, *Astrophys. J. Suppl.* **6**, 167 (1961); D. G. Hummer, *Astrophys. J.* **327**, 472 (1988); R. S. Sutherland, *Mon. Not. Roy. Astron. Soc.* **300**, 321 (1998); P. A. M. van Hoof et al., *Mon. Not. Roy. Astron. Soc.* **444**, 420 (2014).

Equations (3.7.5) and (3.7.6) were given by Karzas and Latter,[37] who derived them from a partial wave expansion of Biedenharn[38] of results originally obtained by Sommerfeld.[39] We will not go through this derivation here. It takes only a glance at Eqs. (3.7.5) and (3.7.6) to see that in order to learn the trend of how the emission depends on frequency and velocity, it would be more convenient to have a simple analytic approximation for the Gaunt factor. This would also make it easy to obtain a simple analytic formula for the thermal Gaunt factor (3.7.4). Above all, in an independent derivation of a simple analytic expression for the Gaunt factor we can gain an understanding of the physics of bremsstrahlung that is not transparent in Eqs. (3.7.5) and (3.7.6).

The most widely useful simple analytic approximation for the Gaunt factor is found using the Born approximation – that is, keeping only terms in the matrix element for bremsstrahlung to first order in the electrostatic interaction between electrons and atoms or ions. This approximation applies if the potential energy, at an electron–nucleus separation r that is equal to the de Broglie wavelengths $\hbar/m_e v$ and $\hbar/m_e v'$ of the initial and final electrons, is much less than the kinetic energies $m_e v^2/2$ or $m_e v'^2/2$. For completely ionized atoms of atomic number Z the potential is $-Ze^2/r$, and the condition for the validity of the Born approximation is

$$\xi \equiv \frac{Ze^2}{\hbar v} \ll 1 \text{ and } \xi' \equiv \frac{Ze^2}{\hbar v'} \ll 1. \qquad (3.7.9)$$

An electron has $\xi < 1$ if its kinetic energy is greater than the binding energy of an atomic electron in the $1s$ state, equal to 13.6 eV for hydrogen. Most of the work needed to deal with bremsstrahlung in the Born approximation has already been done in Section 3.3, where we calculated the bremsstrahlung contribution to cooling. From Eq. (3.3.21) we see that the emission rate per electron is

$$\begin{aligned} j(v, v) &= \frac{4Z^2 e^6 n_N}{3\pi \hbar c^3 m_e^2} \int_0^p \frac{p'^2 \, dp'}{pp'} \ln\left(\frac{p+p'}{p-p'}\right) \delta\left(\frac{p^2}{2m_e h} - \frac{p'^2}{2m_e h} - v\right) \\ &= \frac{8Z^2 e^6 n_N}{3c^3 m_e^2 v} \ln\left(\frac{v+v'}{v-v'}\right), \end{aligned} \qquad (3.7.10)$$

with v' fixed by the energy-conservation condition (3.7.7). Comparing with Eq. (3.7.2), we find that in the Born approximation the Gaunt factor is

$$g_{\text{ff}}(v, v) = \frac{\sqrt{3}}{\pi} \ln\left(\frac{v+v'}{v-v'}\right). \qquad (3.7.11)$$

---

[37] W. J. Karzas and R. Latter, *Astrophys. J. Suppl.* **6**, 167 (1961).
[38] L. C. Biedenharn, *Phys. Rev.* **162**. 262 (1956).
[39] A. J. Sommerfeld, *Atombau und Spektrallinien*, Vol. II (Vieweg & Sohn, Braunschweig, 1939), Chapter 7, Section 5.

3.7 Soft Bremsstrahlung

In particular, for soft photons with $h\nu \ll m_e v^2$, this gives

$$g_{\rm ff}(\nu, v) = \frac{\sqrt{3}}{\pi} \ln\left(\frac{2m_e v^2}{h\nu}\right). \tag{3.7.12}$$

This is in excellent agreement with the numerical results of van Hoof et al.[40] for soft photons with $2h\nu/m_e v^2 \leq 10^{-2}$ and $\xi < 1$. For $h\nu \ll k_B T$ and $Ze^2/\hbar\sqrt{k_B T/m_e} \ll 1$, typical photons are soft and typical electrons have $\xi \ll 1$, and in this case we can use Eq. (3.7.12) in Eq. (3.7.4) and find that the thermal Gaunt factor appearing in Eq. (3.7.3) is

$$\bar{g}_{\rm ff}(\nu, T) = \frac{\sqrt{3}}{\pi}\left[\ln\left(\frac{4k_B T}{h\nu}\right) - \gamma\right], \tag{3.7.13}$$

where γ is the Euler constant, $\gamma = 0.5772157\ldots$.

This is not the result usually given. Several treatises on the interstellar medium[41] give the formula

$$\bar{g}_{\rm ff}(\nu, T) = \frac{\sqrt{3}}{\pi}\left[\ln\left(\frac{(2k_B T)^{3/2}}{\pi Z e^2 \nu m_e^{1/2}}\right) - \frac{5\gamma}{2}\right], \tag{3.7.14}$$

without providing a derivation or an indication of the range of photon frequencies and temperatures in which this formula is valid, other than just that the photons are soft (and, as discussed below, that Debye screening is negligible). Spitzer cites a 1960 calculation of Scheuer[42] for Eq. (3.7.14). Scheuer obtained this result from a calculation of the emission per electron, based entirely on classical scattering theory, which gave the result

$$g_{\rm ff}(\nu, v) = \frac{\sqrt{3}}{\pi}\left[\ln\left(\frac{m_e v^3}{\pi Z e^2 \nu}\right) - \gamma\right] = \frac{\sqrt{3}}{\pi}\left[\ln\left(\frac{2m_e v^2}{h\nu\xi}\right) - \gamma\right]. \tag{3.7.15}$$

As Scheuer found, using this in Eq. (3.7.4) gives the widely quoted thermal Gaunt factor (3.7.14) for soft photons, with $h\nu \ll k_B T$.

It is evident that Eq. (3.7.15) cannot be valid for $\xi \ll 1$, because it does not reduce to the Born-approximation result (3.7.12) in this case, where the Born approximation is known to be valid. Equation (3.7.15) also cannot be valid for a fixed ratio of photon and electron energies and arbitrarily large ξ, where it gives

---

[40] P. A. M. van Hoof et al., *Mon Not. Roy. Astron. Soc.* **444**, 420 (2014).
[41] D. E. Osterbrock, *Astrophysics of Gaseous Nebulae and Active Galactic Nuclei* (University Science Books, Mill Valley, CA, 1989); L. Spitzer, Jr., *Physical Processes in the Interstellar Medium* (John Wiley & Sons, New York, 1998); B. T. Draine, *Physics of the Interstellar and Intergalactic Medium* (Princeton University Press, Princeton, NJ, 2011); W. J. Maciel, *Astrophysics of the Interstellar Medium*, trans. M. Serote Roos (Springer Sciences, New York, 2013).
[42] F. A. G. Scheuer, *Mon. Not. Roy. Astron. Soc.* **120**, 231 (1960). The paper of Scheuer was also quoted as a result of classical scattering theory in an early review article: L. Oster, *Rev. Mod. Phys.* **13**, 525 (1961).

a negative Gaunt factor. It is in fact a good approximation only for $\xi' - \xi \ll 1$,[43] or equivalently, for

$$1 \ll \xi \ll m_e v^2 / 2h\nu. \tag{3.7.16}$$

The numerical results of van Hoof et al.[44] show that the range of values of ξ in which Eq. (3.7.15) is a good approximation is vanishingly narrow even for photon frequencies as low as $10^{-2} m_e v^2 / h$.

It may be possible to go further. We found Eq. (3.7.10) here from the formula (3.3.21) for the Born-approximation emissivity per electron, which followed from a formula, Eq. (3.3.20), for the rate per momentum space volume of the final electron at which an electron of momentum $\mathbf{p}$ is scattered into a definite momentum $\mathbf{p}' \neq \mathbf{p}$ with the emission of a photon of energy $qc = h\nu$. The rate is

$$\sum_\lambda \int \frac{d^2\hat{q}}{4\pi} j^{\text{Born}}(\mathbf{q}, \lambda; \mathbf{p} \to \mathbf{p}') = \frac{2Z^2 e^6 n_N}{3\pi^2 \hbar (\mathbf{p} - \mathbf{p}')^2 q^3 c^3 m_e^2}$$
$$\times \delta\left(\frac{p^2}{2m_e} - \frac{p'^2}{2m_e} - qc\right). \tag{3.7.17}$$

Although the calculation of this photon emission rate in Section 3.3 was based on the Born approximation, which for general frequencies is only valid for electron velocities at which condition (3.7.9) is satisfied, there is a very general soft-photon theorem of quantum electrodynamics,[45] which states that the rate of emission of a photon in any process is given in the limit of vanishing photon energy by a known factor times the rate of the process without the soft photon. In our case the latter process is ordinary Coulomb scattering, whose rate is known to be correctly given by the Born approximation to all orders in the Coulomb potential. Hence from the soft-photon theorem it follows[46] that Eq. (3.7.17) is actually valid to all orders in the Coulomb potential in the limit $qc \to 0$, whether or not condition (3.7.9) is satisfied.

Unfortunately, this is not the end of the story. To derive the emission rate per electron $j(\nu, v)$ from Eq. (3.7.17) we have to integrate over the final electron momentum, as in deriving Eq. (3.3.21) from Eq. (3.3.20). The trouble is that the soft-photon theorem only gives Eq. (3.7.17) in the limit $h\nu = qc = 0$, but in this limit there is a logarithmic divergence in the integral over the direction of the final electron momentum, arising from the configuration in which the final and initial electron momenta are parallel, which accounts for the logarithmic

---

[43] This was found as a consequence of Eqs. (3.7.5) and (3.7.6) by S. Albalat and A. Zimmerman, private communication. They subsequently found that the same condition had been given in an old review article, by P. J. Brussaard and H. C. van de Hulst, *Rev. Mod. Phys.* **34**, 505 (1962).
[44] P. A. M. van Hoof et al. *Mon. Not. Roy. Astron. Soc.* **444**, 420 (2014).
[45] S. Weinberg, *Phys. Rev.* **140**, B516 (1965).
[46] S. Weinberg, *Phys. Rev. D* **99**, 076018 (2019).

dependence of the Gaunt factor on $h\nu$. That is, although Eq. (3.7.17) always holds for emission of photons of sufficiently low energy, the condition on the photon energy for Eq. (3.7.17) to hold becomes increasingly stringent as the final and initial electron directions approach each other. This makes no difference for $Ze^2/\hbar v \ll 1$, where the Born approximation applies anyway, but leads to a suppression of the emissivity otherwise. In order to find an analytic expression for this suppression, it would be necessary to have a good estimate for the angle between initial and final electron velocities within which the emission rate becomes significantly less than the Born-approximation value.

All of the above in this section relies on the assumption that the electron gas is very dilute. For finite electron number density the results (3.7.12) and (3.7.13) are modified by a screening of the ionic charge, known as *Debye screening*,[47] even where the interstellar medium is completely ionized. Because electrons are mobile, their number density in phase space is proportional to

$$\exp\left(-(\mathbf{p}_e^2/2m_e - e\phi(\mathbf{x}))/k_B T\right),$$

where $\phi(\mathbf{x})$ is the electrostatic potential. Integrating over the electron momentum $\mathbf{p}_e$ gives an electron charge density

$$-en_e \exp\left(e\phi(\mathbf{x})/k_B T\right),$$

where n_e is the average electron number density far from specific sources, where $\phi(\mathbf{x}) \simeq 0$. Inserting an ion of charge $+Ze$ at $\mathbf{x} = 0$ then gives an electrostatic potential satisfying the Poisson equation:

$$\nabla^2 \phi(\mathbf{x}) = 4\pi e n_e \exp\left(e\phi(\mathbf{x})/k_B T\right) - 4\pi Z e n_N - 4\pi Z e \delta^3(\mathbf{x}), \quad (3.7.18)$$

in which the second term represents the effect of the average positive ionic charge far from the origin, where $\phi = 0$. The neutrality of the average charge distribution requires that $Zen_N - en_e = 0$, so the first and second terms in Eq. (3.7.18) would cancel if ϕ were zero. Even for $\phi \neq 0$, it is frequently the case that the temperature is high enough to make the argument of the exponential small, in which case we can approximate $\exp\left(e\phi(\mathbf{x})/k_B T\right) \simeq 1 + e\phi(\mathbf{x})/k_B T$, and the Poisson equation becomes

$$\nabla^2 \phi(\mathbf{x}) = \frac{4\pi e^2 n_e}{k_B T} \phi(\mathbf{x}) - 4\pi Z e \delta^3(\mathbf{x}). \quad (3.7.19)$$

This has a well-known solution

$$\phi(\mathbf{x}) = \frac{Ze}{r} \exp(-r/\ell), \quad (3.7.20)$$

---

[47] P. Debye and E. Hückel, *Phys. Z.* **24**, 185 (1923).

where $r \equiv |\mathbf{x}|$ and ℓ is the *Debye length*

$$\ell = \sqrt{\frac{k_B T}{4\pi e^2 n_e}}. \qquad (3.7.21)$$

So at distances beyond ℓ the ionic charge is screened by ambient electrons.

According to Eq. (1.4.A9), the effect of this screening is to replace the denominator $|\mathbf{p} - \mathbf{p}'|^2$ in the matrix element (whether for photon absorption or emission) with $|\mathbf{p} - \mathbf{p}'|^2 + \hbar^2/\ell^2$, so in place of Eq. (3.3.19) the emission per electron in the Born approximation is

$$j^{\text{Born}}(\mathbf{q}, \lambda; \mathbf{p} \to \mathbf{p}') = \frac{Z^2 e^6 n_N}{\pi^2 \hbar [(\mathbf{p} - \mathbf{p}')^2 + \hbar^2/\ell^2]^2 q^3 c^3 m_e^2}$$

$$\times \left| (\mathbf{p} - \mathbf{p}') \cdot e^*(\lambda, \mathbf{q}) \right|^2 \delta\left(\frac{p^2}{2m_e} - \frac{p'^2}{2m_e} - qc \right),$$

from which it follows that

$$j_\nu(T) = \frac{2 Z^2 e^6 n_N}{3\pi^2 \hbar c^3 m_e^2} \int n_e(\mathbf{p}, T) \, d^3 p$$

$$\times \int d^3 p' \frac{|\mathbf{p} - \mathbf{p}'|^2}{[|\mathbf{p} - \mathbf{p}'|^2 + \hbar^2/\ell^2]^2} \delta\left(\frac{p^2}{2m_e h} - \frac{p'^2}{2m_e h} - \nu \right). \qquad (3.7.22)$$

For soft photons with $h\nu \ll k_B T$, the minimum value of $|\mathbf{p} - \mathbf{p}'|^2$ is of order $m_e h^2 \nu^2/k_B T$, so Debye screening has no appreciable effect on soft bremsstrahlung if $h\nu \gg (\hbar/\ell)\sqrt{k_B T/m_e}$, or in other words, if ν is much larger than the plasma frequency $\nu_P \equiv \sqrt{n_e e^2/\pi m_e}$.[48] This is the case, for instance, where interstellar matter has $n_e \simeq 1$ cm^{-3} and $\nu \gg 9$ kHz. In such cases, Debye screening has a negligible effect on soft bremsstrahlung.

On the other hand, where the electron number density is sufficiently large that $\nu \ll \nu_P$, Debye screening provides the effective cut-off for the integral over the final electron direction, and the emissivity (3.7.22) has a well-defined limit for $\nu \to 0$:

$$j_\nu(T) \to \int n_e(\mathbf{p}, T) \, d^3 p \, \frac{4 Z^2 e^6 n_N}{3 c^3 m_e^2 v} \left[\ln(\eta + 1) - \frac{\eta}{\eta + 1} \right], \qquad (3.7.23)$$

where $\eta \equiv 4 m_e v^2 k_B T / h^2 \nu_P^2$. For typical values of ν_P and $k_B T$ we have $\eta \gg 1$, in which case the integral gives a Born-approximation emissivity of form (3.7.3), with thermal Gaunt factor

$$\overline{g}_{\text{ff}}(0, T) = \frac{\sqrt{3}}{\pi} \left[\ln\left(\frac{2^{3/2} k_B T}{h \nu_P} \right) - \frac{1}{2} - \gamma \right]. \qquad (3.7.24)$$

---

[48] The plasma frequency is the frequency of acoustic waves of very long wavelength in a plasma, and it is the frequency below which electromagnetic waves in a plasma are strongly affected by electron–ion attraction.

Bibliography for Chapter 3

- S. F. Dermott, J. H. Hunter, Jr., and R. E. Wilson, eds., *Astrophysical Disks* (New York Academy of Sciences, New York, 1992).
- B. T. Draine, *Physics of the Interstellar and Intergalactic Medium* (Princeton University Press, Princeton, NJ, 2011).
- J. E. Dyson and D. A. Williams, *The Physics of the Interstellar Medium*, 2nd edn. (Institute of Physics Publishing, Bristol, 1997).
- J. Frank, A. R. King, and D. J. Raine, *Accretion Power in Astrophysics* (Cambridge University Press, Cambridge, 1985).
- S. Kato, J. Fukue, and S. Mineshige, *Black Hole Accretion Disks* (Kyoto University Press, Kyoto, 2008).
- W. J. Maciel, *Astrophysics of the Interstellar Medium* (Springer, New York, 2013).
- M. Maggiore, *Gravitational Wave Astronomy* (Oxford University Press, Oxford, 2008 & 2018)
    - Volume 1: *Theory and Experiment*,
    - Volume 2: *Astrophysics and Cosmology.*
- F. Melia, *High Energy Astrophysics* (Princeton University Press, Princeton, NJ, 2009).
- H. Mo, F. van den Bosch, and S. White, *Galaxy Formation and Evolution* (Cambridge University Press, Cambridge, 2010).
- D. E. Osterbrock, *Astrophysics of Gaseous Nebulae and Active Galactic Nuclei* (University Science Books, Mill Valley, CA, 1989).
- L. Spitzer, Jr., *Physical Processes in the Interstellar Medium* (John Wiley & Sons, New York, 1998).
- J. C. Wheeler, ed., *Accretion Disks in Compact Stellar Systems* (World Scientific, Singapore, 1993).

4
Galaxies

In the foregoing chapters we have been chiefly concerned with the stars and interstellar matter in our own galaxy. The observable universe is filled with billions of other galaxies, composed of similar constituents, to which we now turn.

4.1 Collisionless Dynamics

The distribution in position and velocity of stars in a galaxy can be treated by some of the same equilibrium considerations as the distribution of molecules in a gas, but with one great difference: the stars in a galaxy are so far apart that their motion is mostly determined by the gravitational potential of the whole galaxy, rather than by close encounters with individual other stars. The number of stars of any given type in an element d^3x of spatial volume at position $\mathbf{x}$ and in an element d^3v in velocity space at velocity $\mathbf{v}$ is $f(\mathbf{x},\mathbf{v},t)\,d^3x\,d^3v$, where f is the distribution function, given by an averaged sum over stars of this type:

$$f(\mathbf{x},\mathbf{v},t) = \sum_N \overline{\delta^3(\mathbf{x}-\mathbf{x}_N(t))\,\delta^3(\mathbf{v}-\mathbf{v}_N(t))}, \qquad (4.1.1)$$

where $\mathbf{x}_N(t)$ and $\mathbf{v}_N(t)$ are the position and velocity of the Nth star, and the average is over fluctuations of positions and velocities of individual stars (indicated by the bar) during times short compared with the time for the evolution of the galaxy as a whole. The stars' positions and velocities obey the equations of motion

$$\dot{\mathbf{x}}_N(t) = \mathbf{v}_N(t), \qquad \dot{\mathbf{v}}_N(t) = -\nabla\phi(\mathbf{x}_N(t),t), \qquad (4.1.2)$$

where $\phi(\mathbf{x},t)$ is the gravitational potential of the galaxy, due to gas and dark matter as well as stars. Then, taking the time derivative of Eq. (4.1.1), we have the rate of change of the distribution function

$$\frac{\partial}{\partial t}f(\mathbf{x},\mathbf{v},t) = -\mathbf{v}\cdot\nabla_x f(\mathbf{x},\mathbf{v},t) + \nabla_x\phi(\mathbf{x},t)\cdot\nabla_v f(\mathbf{x},\mathbf{v},t). \qquad (4.1.3)$$

This is the *collisionless Boltzmann equation*, which holds except during rare close encounters of stars. It applies separately to stars of each type, though we will not decorate the distribution function with a label distinguishing the different types of star until we need to near the end of this section.

Just as in the kinetic theory of gases, it is often more convenient to work with moments of the distribution function than with the distribution function itself. First, the number density n of stars is simply the integral of the distribution function over velocity:

$$n(\mathbf{x}, t) \equiv \int d^3 v \, f(\mathbf{x}, \mathbf{v}, t). \tag{4.1.4}$$

Assuming that $f(\mathbf{x}, \mathbf{v}, t)$ vanishes rapidly as the velocity goes to infinity, when Eq. (4.1.3) is integrated over velocity the term proportional to $\nabla_v f$ does not contribute to the integral, and we obtain the equation of continuity:

$$\frac{\partial}{\partial t} n(\mathbf{x}, t) + \nabla \cdot \left(n(\mathbf{x}, t) \bar{\mathbf{v}}(\mathbf{x}, t) \right) = 0, \tag{4.1.5}$$

where $\bar{\mathbf{v}}$ is the mean velocity of stars, defined by

$$n(\mathbf{x}, t) \bar{\mathbf{v}}(\mathbf{x}, t) \equiv \int d^3 v \, \mathbf{v} f(\mathbf{x}, \mathbf{v}, t). \tag{4.1.6}$$

Also, multiplying Eq. (4.1.3) with v_i and integrating over velocity gives

$$\frac{\partial}{\partial t} \left(n(\mathbf{x}, t) \bar{v}_i(\mathbf{x}, t) \right) = -\frac{\partial}{\partial x_j} \left(n(\mathbf{x}, t) \overline{v_i v_j}(\mathbf{x}, t) \right) - n(\mathbf{x}, t) \frac{\partial}{\partial x_i} \phi(\mathbf{x}, t), \tag{4.1.7}$$

where $\overline{v_i v_j}$ is defined by

$$n(\mathbf{x}, t) \overline{v_i v_j}(\mathbf{x}, t) \equiv \int d^3 v \, v_i v_j f(\mathbf{x}, \mathbf{v}, t).$$

(The third term in Eq. (4.1.3) has here been integrated by parts.)

Equation (4.1.7) has a nice application[1] to the motion of stars in the flat disks of galaxies like our own and M31. We can assume that the distribution of stars is in equilibrium, so that we can drop the time derivative in Eq. (4.1.7), and that the distribution depends much less on location within the area of the disk than on the distance z from the disk's mid-plane along a direction normal to the disk. Then, taking the z-component of Eq. (4.1.7), and assuming that the quantities involved depend chiefly on z, we find

$$\frac{1}{n} \frac{d}{dz} \left(n \overline{v_z^2} \right) = -\frac{d}{dz} \phi.$$

---

[1] J. H. Oort, *Bull. Astron. Inst. Netherlands* **6**, 349 (1932).

With ϕ depending chiefly on z, the mass density ρ can be found from the Poisson equation:

$$4\pi G\rho = \nabla^2\phi = \frac{d^2\phi}{dz^2} = -\frac{d}{dz}\left(\frac{1}{n}\frac{d}{dz}\left(n\overline{v_z^2}\right)\right). \qquad (4.1.8)$$

Note that in deriving the right-hand side we used only the Boltzmann equation and the equation of motion, which apply separately for each type of star, and in particular the right-hand side is independent of the overall scale of n. Hence we can take n here to be the number density of any easily observable test bodies that move under the influence of the galaxy's gravitational field, such as bright stars, whether or not they make an appreciable contribution to the galaxy's mass density, which may arise largely from invisible interstellar matter. By evaluating the right-hand side, astronomers have worked out[2] that the total mass density in the mid-plane of our galaxy is about 10^{-23} g/cm^3. The mass density of observed stars is about 4×10^{-24} g/cm^3, the missing mass presumably consisting of dark matter and/or stars too faint to observe. This calculation is often used to provide an upper limit on the density of such faint stars, and is referred to as the *Oort limit*.

Returning now to general collisionless dynamics, we can use Eq. (4.1.5) to rewrite Eq. (4.1.7) in the useful form

$$\frac{\partial}{\partial t}\overline{v}_i(\mathbf{x},t) = -(\overline{\mathbf{v}}(\mathbf{x},t)\cdot\nabla)\overline{v}_i(\mathbf{x},t) - \frac{1}{n(\mathbf{x},t)}\frac{\partial}{\partial x_j}\Pi_{ij}(\mathbf{x},t) - \frac{\partial}{\partial x_i}\phi(\mathbf{x},t), \quad (4.1.9)$$

where Π_{ij} is the velocity dispersion tensor

$$\begin{aligned}\Pi_{ij}(\mathbf{x},t) &\equiv n(\mathbf{x},t)\left(\overline{v_i(\mathbf{x},t)v_j(\mathbf{x},t)} - \overline{v}_i(\mathbf{x},t)\overline{v}_j(\mathbf{x},t)\right) \\ &= \int d^3v\left(v_iv_j - \overline{v}_i(\mathbf{x},t)\overline{v}_j(\mathbf{x},t)\right)f(\mathbf{x},\mathbf{v},t) \\ &= \int d^3v\left(v_i - \overline{v}_i(\mathbf{x},t)\right)\left(v_j - \overline{v}_j(\mathbf{x},t)\right)f(\mathbf{x},\mathbf{v},t). \qquad (4.1.10)\end{aligned}$$

If we liked we could obtain a formula for the rate of change of Π_{ij} by multiplying Eq. (4.1.3) with v_iv_j and integrating over $\mathbf{v}$, but this would be a losing game; the formula derived in this way would involve the average of a product of *three* velocity components. This process never comes to an end. Instead, it is often sufficient to provide closure by assuming that (perhaps because of the rare close encounters of stars) the distribution of $\mathbf{v} - \overline{\mathbf{v}}$, the stellar velocity relative to its mean value, is isotropic, in which case Π_{ij} takes the form

$$\Pi_{ij}(\mathbf{x},t) = \delta_{ij}\Pi(\mathbf{x},t), \qquad (4.1.11)$$

---

[2] L. Spitzer, Jr., *Physical Processes in the Interstellar Medium* (John Wiley & Sons, New York, 1978), Section 1.6.

and Eq. (4.1.9) becomes

$$\frac{\partial}{\partial t}\bar{\mathbf{v}}(\mathbf{x},t) = -\big(\bar{\mathbf{v}}(\mathbf{x},t)\cdot\nabla\big)\bar{\mathbf{v}}(\mathbf{x},t) - \frac{1}{n(\mathbf{x},t)}\nabla\Pi(\mathbf{x},t) - \nabla\phi(\mathbf{x},t). \qquad (4.1.12)$$

This is pretty much the same as the Euler equation (3.4.12) of hydrodynamics in the presence of a gravitational potential, but with Π and n playing the roles of pressure and mass density. Of course, we still need some way of estimating Π as a function of other dynamical variables.

We will be primarily interested in the condition for equilibrium, in which f does not depend on time:

$$\frac{\partial}{\partial t}f(\mathbf{x},\mathbf{v},t) = 0. \qquad (4.1.13)$$

Using Eq. (4.1.3) and dropping the time argument, this is

$$0 = -\mathbf{v}\cdot\nabla_x f(\mathbf{x},\mathbf{v}) - \nabla\phi(\mathbf{x})\cdot\nabla_v f(\mathbf{x},\mathbf{v}). \qquad (4.1.14)$$

It is not hard to find a large class of solutions of Eq. (4.1.14). Suppose $I_r(\mathbf{x},\mathbf{v})$ are various integrals of the motion, in the sense that the $I_r(\mathbf{x}(t),\mathbf{v}(t))$ are time-independent for any phase-space trajectory satisfying Eqs. (4.1.2):

$$\dot{\mathbf{x}}(t) = \mathbf{v}(t), \qquad \dot{\mathbf{v}}(t) = -\nabla\phi\big(\mathbf{x}(t)\big). \qquad (4.1.15)$$

An obvious integral of the motion is the energy per mass

$$E(\mathbf{x},\mathbf{v}) = \frac{1}{2}\mathbf{v}^2 + \phi(\mathbf{x}). \qquad (4.1.16)$$

If the distribution function depends only on $E(\mathbf{x},\mathbf{v})$, then it is an isotropic function of $\mathbf{v}$, and satifies Eqs. (4.1.11) and (4.1.12). The existence of other possible integrals of the motion depends on possible symmetries of the galaxy. In galaxies with symmetry about an axis of rotation, the angular momentum J about that axis is another such integral.

Any distribution function of the form

$$f(\mathbf{x},\mathbf{v},t) = \mathcal{F}\big(I(\mathbf{x}(t),\mathbf{v}(t))\big) \qquad (4.1.17)$$

will satisfy

$$\frac{\partial}{\partial t}f(\mathbf{x},\mathbf{v},t) = \sum_r \frac{\partial \mathcal{F}}{\partial I_r}\frac{d}{dt}I_r(\mathbf{x}(t),\mathbf{v})t)) = 0, \qquad (4.1.18)$$

so that (4.1.17) is a solution of the equilibrium equation (4.1.13). Since f and I_r are thus independent of time, we can drop the time argument in Eq. (4.1.17), which reads

$$f(\mathbf{x},\mathbf{v}) = \mathcal{F}\big(I(\mathbf{x},\mathbf{v})\big). \qquad (4.1.19)$$

From Eq. (4.1.18) it follows that this is a solution of the equilibrium Boltzmann equation (4.1.14), a result known as the *Jeans theorem*. We will see examples of this procedure in the next section.

Let's look in more detail at the simplest case of Eq. (4.1.17), in which the distribution function $f(\mathbf{x}, \mathbf{v})$ depends on a single conserved quantity, the stellar energy per mass (4.1.16):

$$f(\mathbf{x}, \mathbf{v}) = \mathcal{F}(\mathbf{v}^2/2 + \phi(\mathbf{x})).$$

This is a case in which the mean velocity $\bar{v}$ defined by Eq. (4.1.6) is everywhere zero, so it does not apply to stars in galaxies with net rotation, like the disk galaxies considered in Section 4.3, but it allows a fair account of spherical or ellipsoidal galaxies. We can write a formula for the equilibrium number density as an integral over E

$$n(\mathbf{x}) = \mathcal{N}(\phi(\mathbf{x})), \qquad (4.1.20)$$

with

$$\mathcal{N}(\phi) = \int d^3v \, \mathcal{F}(\mathbf{v}^2/2 + \phi) = 4\pi \int_\phi^\infty dE \sqrt{2(E-\phi)} \mathcal{F}(E). \qquad (4.1.21)$$

To avoid problems with the limits of integration, it is usually assumed that $\mathcal{F}(E)$ goes to zero smoothly at a maximum energy $E = E_{\max}$, and vanishes for all $E > E_{\max}$, so that Eq. (4.1.21) reads

$$\mathcal{N}(\phi) = 4\pi \int_\phi^{E_{\max}} dE \sqrt{2(E-\phi)} \mathcal{F}(E). \qquad (4.1.22)$$

It follows from this that $\mathcal{N}(\phi)$ vanishes for $\phi > E_{\max}$, though of course it may vanish also for smaller values of ϕ. In typical cases $\phi(\mathbf{x})$ is everywhere negative, and we can take $E_{\max} = 0$.

This can be inverted. Not only can we find for every function $\mathcal{F}(E)$ an equilibrium number density distribution (4.1.21) that depends only on ϕ; it is also the case that for any number density distribution of the form (4.1.20) that depends only on ϕ and that vanishes for ϕ greater than some maximum $E_{\max}$, we can find an equilibrium distribution function of the form $f(\mathbf{x}, \mathbf{v}) = \mathcal{F}(E(\mathbf{x}, \mathbf{v}))$ for which $\mathcal{N}(\phi)$ is given by Eq. (4.1.22). This distribution function is

$$\mathcal{F}(E) = \frac{1}{\sqrt{8}\pi^2} \frac{d}{dE} \int_E^{E_{\max}} \frac{d\mathcal{N}(\psi)}{d\psi} \frac{d\psi}{\sqrt{\psi - E}}. \qquad (4.1.23)$$

(This is a generalization of what is sometimes called the *Eddington formula*.[3]) To check this, note that if we use Eq. (4.1.23) in the right-hand side of Eq. (4.1.22), we have

$$4\pi \int_\phi^{E_{\max}} dE \sqrt{2(E-\phi)} \mathcal{F}(E) = \frac{2}{\pi} \int_\phi^{E_{\max}} dE \sqrt{E-\phi}$$
$$\times \frac{d}{dE} \int_E^{E_{\max}} \frac{d\mathcal{N}(\psi)}{d\psi} \frac{d\psi}{\sqrt{\psi - E}}.$$

---
[3] A. S. Eddington, *Mon. Not. Roy. Astron. Soc.* **76**, 572 (1916).

The integral over ψ vanishes at the upper endpoint $E = E_{\max}$ of the integral over E, and $\sqrt{E-\phi}$ vanishes at the lower endpoint $E = \phi$ of the integral over E, so integrating over E by parts gives

$$4\pi \int_\phi^{E_{\max}} dE \sqrt{2(E-\phi)} \mathcal{F}(E) = -\frac{1}{\pi} \int_\phi^{E_{\max}} \frac{dE}{\sqrt{E-\phi}} \int_E^{E_{\max}} \frac{d\mathcal{N}(\psi)}{d\psi} \frac{d\psi}{\sqrt{\psi - E}}.$$

Inverting the order of integrations, this is

$$4\pi \int_\phi^{E_{\max}} dE \sqrt{2(E-\phi)} \mathcal{F}(E) = -\frac{1}{\pi} \int_\phi^{E_{\max}} d\psi \frac{d\mathcal{N}(\psi)}{d\psi} \int_\phi^\psi \frac{dE}{\sqrt{E-\phi}\sqrt{\psi - E}}.$$

The integral over E is independent of ϕ and ψ:

$$\int_\phi^\psi \frac{dE}{\sqrt{E-\phi}\sqrt{\psi-E}} = \pi,$$

so

$$4\pi \int_\phi^{E_{\max}} dE \sqrt{2(E-\phi)} \mathcal{F}(E) = -\int_\phi^{E_{\max}} d\psi \frac{d\mathcal{N}(\psi)}{d\psi} = \mathcal{N}(\phi)$$

as was to be shown.

Despite this demonstration, Eqs. (4.1.23) and (4.1.19) do not quite provide a universally applicable prescription for constructing an equilibrium distribution function $f(\mathbf{x}, \mathbf{v})$ for any given density profile $n(\mathbf{x})$. For one thing, Eq. (4.1.20) works only if there is a one-to-one correspondence between values of $\mathbf{x}$ and of $\phi(\mathbf{x})$. This is the case, for instance, for a spherically symmetric mass distribution, such that as usual $\phi(r)$ increases monotonically with increasing r (or, since ϕ is negative, $|\phi(r)|$ decreases monotonically with r). Another limitation is that Eq. (4.1.23) makes sense only if it gives a distribution function $\mathcal{F}(E)$ that is everywhere positive, which is not guaranteed. The above argument based on Eq. (4.1.23) only shows that it does not take a miracle for a given density profile to result from some distribution function of the form $\mathcal{F}(E)$.

For a distribution function $f(\mathbf{x}, \mathbf{v}) = \mathcal{F}(\mathbf{v}^2/2 + \phi(\mathbf{x}))$, the tensor $\Pi_{ij}(\mathbf{x})$ defined by Eq. (4.1.10) obviously has the form $\Pi_{ij} = \delta_{ij} \Pi$ suggested in Eq. (4.1.11), so these stars behave as a fluid, with a mass density and pressure given by

$$\rho(\mathbf{x}) = \sum_s m_s n_s(\mathbf{x}) = \sum_s m_s \int d^3v \, \mathcal{F}_s(\mathbf{v}^2/2 + \phi(\mathbf{x})), \quad (4.1.24)$$

$$p(\mathbf{x}) \equiv \sum_s m_s \Pi_s(\mathbf{x}) = \sum_s \frac{m_s}{3} \int d^3v \, \mathbf{v}^2 \mathcal{F}_s(\mathbf{v}^2/2 + \phi(\mathbf{x})), \quad (4.1.25)$$

the sums now running over all types s of star, each type taken to have a common mass m_s. Since the distribution functions $\mathcal{F}_s(\mathbf{v}^2/2 + \phi(\mathbf{x}))$ automatically satisfy the time-independent Boltzmann equation, we expect the pressure and density automatically to satisfy the equation of hydrostatic equilibrium, whatever the form of $\mathcal{F}_s$. To check this, note that

$$\nabla p = \sum_s \left(\frac{m_s}{3}\right) \nabla \phi \int d^3v \, \mathbf{v}^2 \mathcal{F}'_s(\mathbf{v}^2/2 + \phi(\mathbf{x})) \, .$$

Note also that

$$(\mathbf{v} \cdot \nabla_v)\mathcal{F}_s(\mathbf{v}^2/2 + \phi(\mathbf{x})) = \mathbf{v}^2 \mathcal{F}'_s(\mathbf{v}^2/2 + \phi(\mathbf{x})),$$

so

$$\nabla p = \sum_s \left(\frac{m_s}{3}\right) \nabla \phi \int d^3v \, (\mathbf{v} \cdot \nabla_v)\mathcal{F}_s(\mathbf{v}^2/2 + \phi(\mathbf{x})).$$

Integrating by parts and using $\nabla_v \cdot \mathbf{v} = 3$, we obtain the general equation of hydrostatic equilibrium

$$\nabla p = -\rho \nabla \phi,$$

as was to be shown. But we do not complete the work of constructing a galaxy model just by choosing functions $\mathcal{F}_s(E)$, for we still have the task of finding $\phi(\mathbf{x})$ by solving the Poisson equation

$$\nabla^2 \phi(\mathbf{x}) = 4\pi G \sum_s m_s \int d^3v \, \mathcal{F}_s(\mathbf{v}^2/2 + \phi(\mathbf{x})), \qquad (4.1.26)$$

which is not so easy for general $\mathcal{F}_s$. In the next section we will see how this can be done for some special choices of the distribution functions $\mathcal{F}_s$.

The same formalism can be applied in other contexts. It is not only stars in a galaxy that interact chiefly with the gravitational potential of the whole galaxy. Because the interactions of dark matter particles are believed to be so weak, their motion likewise is governed by the gravitational field of the galaxy or cluster of galaxies they inhabit. Likewise, in clusters of galaxies, the individual galaxies are so far apart that they interact chiefly with the gravitational potential of the cluster. In all these cases, the formalism presented above allows the construction of equilibrium distribution functions for galaxies in clusters and for dark matter particles in galaxies or clusters.

4.2 Polytropes and Isothermals

We can use the general method for finding solutions of the equilibrium Boltzmann equation described in the previous section to construct plausible models of spherical halos, galaxies, and galaxy clusters. We take the equilibrium distribution function for bodies (particles, stars, or galaxies) of type s to have the

4.2 Polytropes and Isothermals

form (4.1.19), with a single invariant I in this equation chosen as the energy per mass (4.1.16):

$$f_s(\mathbf{x}, \mathbf{v}) = \mathcal{F}_s(\mathbf{v}^2/2 + \phi(\mathbf{x})), \tag{4.2.1}$$

and now choose the functions $\mathcal{F}_s$ to have the simple form

$$\mathcal{F}_s(E) = \begin{cases} C_s(E_m - E)^\nu & E \leq E_m \\ 0 & E \geq E_m, \end{cases} \tag{4.2.2}$$

where $C_s > 0$, $\nu \geq 0$, and E_m are constants, with E_m at least equal to the maximum value reached by $\phi(\mathbf{x})$. It will be a great convenience to take ν and E_m to be the same for all s, though C_s may vary from one type of body to another.

This is of course a large oversimplification, but it is not a bad approximation in some contexts, such as the outer parts of spheroidal galaxies. More generally we could have a different $\mathcal{F}(E)$, with different values for C, ν, and E_m for each type of star (or each type of galaxy in a cluster, or each type of dark matter particle in a halo, galaxy, or cluster). But of course in any case $\phi(\mathbf{x})$ is the same for all s, given by a sum of contributions of stars, gas, and dark matter.

The mass density and pressure are given by the integrals

$$\rho(\mathbf{x}) = \sum_s m_s \int d^3v \, \mathcal{F}_s(\mathbf{v}^2/2 + \phi(\mathbf{x})) \tag{4.2.3}$$

and

$$p(\mathbf{x}) = \sum_s \frac{m_s}{3} \int d^3v \, \mathbf{v}^2 \mathcal{F}_s(\mathbf{v}^2/2 + \phi(\mathbf{x})), \tag{4.2.4}$$

where, depending on the context, the sums run over types of star in a galaxy, types of galaxies in a cluster of galaxies, or types of dark matter particle in either, with m_s taken as the common mass of each type. The integrals can easily be done by changing the variable of integration from v to $u \equiv v^2[2(E_m - \phi(\mathbf{x}))]^{-1}$, and using the general formula[4]

$$\int_0^1 du \, u^\mu (1-u)^\nu = \frac{\Gamma(\nu+1)\Gamma(\mu+1)}{\Gamma(\mu+\nu+2)}.$$

This gives the density and pressure

$$\rho(\mathbf{x}) = (2\pi)^{3/2} (E_m - \phi(\mathbf{x}))^{\nu+3/2} \frac{\Gamma(\nu+1)}{\Gamma(\nu+5/2)} \sum_s m_s C_s \tag{4.2.5}$$

and

$$p(\mathbf{x}) = (2\pi)^{3/2} (E_m - \phi(\mathbf{x}))^{\nu+5/2} \frac{\Gamma(\nu+1)}{\Gamma(\nu+7/2)} \sum_s m_s C_s. \tag{4.2.6}$$

---

[4] Here $\Gamma(z)$ is the familiar Gamma function, and of course has no connection with the polytrope index Γ calculated below. We shall use the formulas $\Gamma(3/2) = \sqrt{\pi}/2$ and $\Gamma(5/2) = 3\sqrt{\pi}/4$.

We now specialize to the case of a spherically symmetric potential, for which ϕ and consequently ρ and p are all functions only of the radial coordinate r. As remarked in the previous section, because the distribution function (4.2.1) automatically satisfies the equilibrium Boltzmann equation, the density and pressure we have calculated must automatically satisfy the equation of hydrostatic equilibrium. This can easily be checked here. Using the formula

$$\frac{\nu + 5/2}{\Gamma(\nu + 7/2)} = \frac{1}{\Gamma(\nu + 5/2)},$$

Eqs. (4.2.5) and (4.2.6) give

$$\frac{dp}{dr} = -\rho \frac{d\phi}{dr},$$

which with the Poisson equation is the same as Eq. (1.1.4). But Eqs. (4.2.5) and (4.2.6) do not end the need for solving difficult differential equations, because we still have to calculate ϕ by solving the Poisson equation (4.1.26), which here reads

$$\frac{1}{r^2}\frac{d}{dr}\left(r^2\frac{d\phi}{dr}\right) = 4\pi G\rho = 4\pi G(2\pi)^{3/2}(E_m - \phi)^{\nu+3/2}\frac{\Gamma(\nu+1)}{\Gamma(\nu+5/2)}\sum_s m_s C_s.$$

It helps in this task to note that because we take ν to be the same for each type of star (or galaxy, or particle), we have the same relation here between pressure and density as for the polytropes discussed in Section 1.8:

$$p = K\rho^\Gamma, \tag{4.2.7}$$

where here

$$\Gamma = \frac{\nu + 5/2}{\nu + 3/2} \tag{4.2.8}$$

and

$$K = \left((2\pi)^{3/2}\sum_s m_s C_s \Gamma(\nu+1)\right)^{\Gamma-1}\frac{\Gamma(\nu+5/2)^{\Gamma-1}}{\Gamma(\nu+7/2)}. \tag{4.2.9}$$

As we learned in Section 1.8, the density and pressure for such a polytrope drop to zero at a finite radius R, as long as $\Gamma > 6/5$. For non-singular distribution functions (4.2.2) we have $\nu \geq 0$, so Eq. (4.2.8) gives a range of values for the polytrope index

$$1 < \Gamma \leq 5/3, \tag{4.2.10}$$

which includes both density distributions with finite radii and those that extend to infinity.

There is special importance to the polytropes with $\Gamma = 1$, for which $p(r)$ is simply proportional to $\rho(r)$. This is like a gas with an r-independent temperature, and hence is called an *isothermal distribution*. We cannot quite get

4.2 Polytropes and Isothermals

down to $\Gamma = 1$ with a distribution function of the form (4.2.2), but instead we can take

$$\mathcal{F}_s(E) = A_s \exp(-BE), \tag{4.2.11}$$

with A_s and B positive constants. We take B the same for all s because it here plays something like the role of temperature, although since $E = \mathbf{v}^2/2 + \phi$ is not the energy but the energy per mass, B does not quite correspond to the inverse temperature $1/k_B T$ in energy units, but instead corresponds to the ratio of the stellar mass m_s to $k_B T$. We are here giving up the strict requirement that all stars are in bound orbits with $E < 0$, but just as for molecules of oxygen and nitrogen in the Earth's atmosphere, if B is sufficiently large the evaporation of stars from the galaxy will be very slow.

The mass density and pressure for the distribution (4.2.11) are

$$\rho(\mathbf{x}) = \sum_s m_s \int d^3v \, f_s(\mathbf{x}, \mathbf{v}) = (2\pi/B)^{3/2} \sum_s A_s m_s \exp\left(-B\phi(\mathbf{x})\right) \tag{4.2.12}$$

and

$$p(\mathbf{x}) = \sum_s \frac{m_s}{3} \int d^3v \, v^2 f_s(\mathbf{x}, \mathbf{v}) = \frac{\rho(\mathbf{x})}{B}. \tag{4.2.13}$$

For spherical symmetry, with $p(r) = \rho(r)/B$, this is what in Section 1.8 was called a polytrope, here with $\Gamma = 1$ and $K = 1/B$. Because $\rho \propto \exp(-B\phi)$, there is no surface at which ρ drops to zero, as also follows from the fact that this is a polytrope with $\Gamma < 6/5$. For this reason, Eq. (4.2.11) cannot describe an actual spherical galaxy or galaxy cluster throughout its volume, though as we shall see, it may apply to the inner parts of some galaxies and clusters of galaxies.

Here again, because the $\mathcal{F}_s$ are invariant under rotations of $\mathbf{v}$ alone, and are time-independent, ρ and p automatically obey the equation of hydrostatic equilibrium. Using Eqs. (4.2.12) and (4.2.13), we have

$$\frac{d}{dr}p(r) = \frac{1}{B}\frac{d}{dr}\rho(r) = -\rho(r)\frac{d}{dr}\phi(r) = -G\rho(r)\mathcal{M}(r)/r^2, \tag{4.2.14}$$

which is the same as Eq. (1.1.4). This can be used with Eq. (4.2.13) and the definition of $\mathcal{M}$ to derive the Lane–Emden equation for a $\Gamma = 1$ polytrope:

$$\frac{d}{dr}\left(\frac{r^2}{\rho}\frac{d\rho}{dr}\right) + 4\pi G B r^2 \rho = 0. \tag{4.2.15}$$

Surprisingly, this has a simple analytic solution

$$\rho(r) = \frac{1}{2\pi G B r^2}. \tag{4.2.16}$$

This is unrealistic at $r = 0$ as well as at $r \to \infty$, but it provides a useful guide to an approximate solution of Eq. (4.2.15) that behaves reasonably at least for finite r, including $r = 0$. For this purpose, let us convert Eq. (4.2.15) to a parameter-free form by the same sort of re-scaling as we used in Section 1.8. Suppose we define a new coordinate and a new density function:

$$F(z) \equiv \rho(r)/\rho(0), \quad z \equiv r\sqrt{4\pi G \rho(0) B}. \tag{4.2.17}$$

Equation (4.2.15) then takes the form

$$\frac{d}{dz}\left(\frac{z^2}{F(z)}\frac{dF(z)}{dz}\right) + z^2 F(z) = 0. \tag{4.2.18}$$

Because $\rho(\mathbf{x})$ must be analytic in $\mathbf{x}$, for $z \to 0$ we must have $F(z) \to 1 + az^2$. Putting this in Eq. (4.2.18) gives $a = -1/6$, so for $z \to 0$,

$$F(z) \to 1 - z^2/6. \tag{4.2.19}$$

On the other hand, for $z \to \infty$

$$F(z) \to 2/z^2, \tag{4.2.20}$$

corresponding to Eq. (4.2.16).[5]

The transition between these two limiting forms obviously occurs for z of order unity, corresponding to a core radius

$$r_c \equiv (4\pi G \rho(0) B)^{-1/2}. \tag{4.2.21}$$

Outside the core $\rho(r)$ becomes proportional to $1/r^2$, so $\mathcal{M}(r) \propto r$, and for circular orbits Newton's relation $v_\theta^2(r)/r \propto \mathcal{M}(r)/r^2$ gives $v_\theta(r)$ constant. As discussed in the next section, r-independent stellar velocities are indeed observed in the outer parts of disk galaxies, though this in itself does not prove that these galaxies have an isothermal halo.

Clusters of galaxies contain two kinds of gas: ordinary baryonic matter, mostly hot ionized hydrogen and helium; and dark matter, whose presence is known only through its contribution to the gravitational field. It is generally assumed that the same "violent relaxation" (close encounters whose gravitational effects cannot be represented as an interaction with a smooth intracluster gravitational field) that causes the concentration of the hot baryonic gas in the cluster is also responsible for the concentration of dark matter, so that baryonic matter and dark matter have the same distribution in E, and hence the same value of B. In this case, Eq. (4.2.14) applies to both the baryonic and dark matter densities $\rho_B(r)$ and $\rho_D(r)$, and can be written

---

[5] To check that $F(z)$ approaches the exact solution $2/z^2$, consider the perturbation to this solution, $F(z) = 2/z^2 + \epsilon(z)$, with $\epsilon(z)$ very small. The term in Eq. (4.2.18) of first order in $\epsilon(z)$ is

$$z^2 \epsilon''(z) + 6z\epsilon'(z) + 8\epsilon(z) = 0.$$

The general solution is $\epsilon(z) \propto z^{-5/2} \cos\left(\sqrt{7}z/2 + \delta\right)$ (with δ an arbitrary phase) which decays as $z \to \infty$. This does not actually prove that $F(z) \to 2/z^2$ for $z \to \infty$, but only that if the function $F(z)$ ever gets close to $2/z^2$ it keeps getting closer as $z \to \infty$.

4.2 Polytropes and Isothermals

$$\frac{d}{dr}\rho_B(r) = -GB\rho_B(r)\mathcal{M}(r)/r^2, \quad (4.2.22)$$

$$\frac{d}{dr}\rho_D(r) = -GB\rho_D(r)\mathcal{M}(r)/r^2, \quad (4.2.23)$$

where $\mathcal{M}(r)$ is the *total* mass within a sphere of radius r:

$$\mathcal{M}(r) = \int_0^r 4\pi r'^2 \rho_M(r')\,dr'; \qquad \rho_M(r) \equiv \rho_B(r) + \rho_D(r). \quad (4.2.24)$$

Adding Eqs. (4.2.22) and (4.2.23) gives

$$\frac{d}{dr}\rho_M(r) = -GB\rho_M(r)\mathcal{M}(r)/r^2, \quad (4.2.25)$$

so Eqs. (4.2.15)–(4.2.22) all apply to $\rho_M(r)$. In particular,

$$\rho_M(r) = \rho_M(0)F(z), \quad z = r/r_c, \quad r_c = (4\pi G\rho_M(0)B)^{-1/2}, \quad (4.2.26)$$

with the function $F(z)$ the same as before. The solutions of Eqs. (4.2.22) and (4.2.23) are then

$$\rho_B(r) = \rho_B(0)F(r/r_c), \qquad \rho_D(r) = \rho_D(0)F(r/r_c). \quad (4.2.27)$$

The parameter B can be found from observation of the distribution of baryonic velocities; the total central density $\rho_M(0)$ can be found from observation of the core radius r_c; and the baryonic central density $\rho_B(0)$ can be found from observation of the X-ray luminosity L_X. Since X-rays are emitted in the collisions of electrons with ions, the X-ray energy per volume is proportional to the square of the baryon density, and so can be written as $\Lambda_X(B)\rho_B^2$. Then the total X-ray luminosity is

$$L_X = \Lambda_X(B)\int_0^R 4\pi r^2 \rho_B^2(r)\,dr = 4\pi\mathcal{I}\Lambda_X(B)r_c^3\rho_B^2(0), \quad (4.2.28)$$

where

$$\mathcal{I} \equiv \int_0^\infty F^2(z)z^2\,dz = 0.1961. \quad (4.2.29)$$

(Recall that $F(z) \to 2/z^2$ for $z \to \infty$, so large values of z are suppressed in $\mathcal{I}$, and the isothermal distribution gives a finite X-ray luminosity, close to the actual value.)

With these assumptions, the ratio of $\rho_B(0)$ to $\rho_M(0)$ found in this way would give the ratio of densities of baryonic mass to all mass throughout the intracluster gas. For decades, this suggested a discrepancy between the ratio of baryonic mass to all mass in galaxy clusters and the same ratio found for the whole universe from observations of anisotropies in the cosmic microwave background.

For instance, a typical reported value[6] for this ratio in clusters is $\rho_B/\rho_M \approx$ 0.12 to 0.13. (Galaxies are of course mostly baryonic, because only baryonic matter can cool and hence condense into galaxies, but galaxies contribute only a small part of the mass of galaxy clusters.) In 2006 X-ray observations of 13 galaxy clusters[7] showed that the baryonic fraction rises to 0.10 to 0.15 toward the outer parts of the cluster, about 1,000 kpc from the center. On the other hand, WMAP observations of microwave background anisotropies gave a value[8] about 0.175 for the baryon/total mass ratio for the whole universe. More recently, observations of the cosmic microwave background with the Planck satellite[9] have lowered the estimated cosmic baryon/total mass ratio to 0.157 ± 0.004, but a recent study[10] of 91 galaxy clusters with redshifts z from 0.2 to 1.25 reports that the baryon/total mass density found in clusters is still less than the Planck result, now by $18 \pm 2\%$.

On the other hand, the assumptions leading to Eqs. (4.2.26)–(4.2.29) need to be modified because shock heating drives baryonic gas away from the inner parts of clusters, while leaving dark matter in place. It is therefore only in the outer parts of clusters that we expect to find ratios of baryonic mass to all mass in agreement with the cosmic ratio. Observed X-rays mostly come from the inner parts of clusters, which biases measurements of these ratios to lower values. An extrapolation of these measurements to the outer parts of clusters indicates that in these parts the ratio of baryonic mass to all mass reaches its cosmic value.[11] A similar conclusion has been reported for the halos of individual galaxies.[12]

Computer simulations of the evolution of dark matter halos give a more complicated picture of the density profile of dark matter. These simulations suggest an essentially universal "NFW" distribution:[13]

$$\rho(r) \propto r^{-1}(1 + r/r_1)^{-2},$$

where r_1 is an adjustable parameter. Unlike the polytrope isothermal solution (4.2.17) this is singular at $r = 0$, though less singular than the simple isothermal solution (4.2.16), and it decreases more rapidly as $r \to \infty$ than either isothermal solution, though still not fast enough to give a finite total mass if not truncated. It gives velocities $v_\theta(r)$ that are roughly constant only for an intermediate range of r, the velocities increasing and decreasing with r for smaller and larger r, respectively. It is still expected that the baryonic gas in clusters of galaxies is approximately isothermal, but, with the gravitational field

---

[6] S. Schindler, *Space Sci. Rev.* **100**, 299 (2002) [astro-ph/0107028].
[7] A. Vikhlinin et al., *Astrophys. J.* **640**, 691 (2006).
[8] D. N. Spergel et al., *Astrophys. J. Suppl.* **170**, 177 (2007) [astro-ph/0603449].
[9] Planck Collective XIII, *Astron. Astrophys.* **594**, A13 (2016).
[10] I. Chiu et al., *Mon. Not. Roy. Astron. Soc.* **478**, 3072 (2018).
[11] B. Rasheed, N. Bahcall, and P. Bode, *Proc. Nat. Acad. Sci.* **108**, 3487 (2011).
[12] J. N. Bregman et al., *Astrophys. J.* **862**, no. 1, paper 3 (2018).
[13] J. F. Navarro, C. S. Frenk, and S. D. M. White, *Mon. Not. Roy. Astron. Soc.* **275**, 56, 270 (1995); J. F. Navarro, C. S. Frenk, and S. D. M. White, *Astrophys. J.* **462**, 563 (1996).

dominated by dark matter, the departures of dark matter from an isothermal distribution lead to differences in the shapes of the baryonic and dark matter density profiles.

4.3 Galactic Disks

Disk galaxies are among the most beautiful astronomical objects revealed by telescopic observation. They typically have a large central bulge, surrounded by a thin rotating disk, decorated with spiral arms to be discussed in the following section.

There are three main contributions to the gravitational potential in the plane of the galaxy, which dominate the potential at different locations. In order of distance from the center, the dominant contributions are the central bulge, the disk itself, and a spherical halo of mostly dark matter. The potential is mapped out by measuring velocities of stars, clusters, masers, etc.[14] For cylindrical symmetry, the tangential component $v_\theta(R)$ of the velocity in the plane of the galaxy at a distance R from the center is related to the potential $\phi(R)$ by the Newtonian relation

$$\frac{v_\theta^2(R)}{R} = \frac{d\phi(R)}{dR}. \tag{4.3.1}$$

For R less than about 1 kpc the potential is entirely dominated by the bulge, so for spherical symmetry we have $\phi'(R) = G\mathcal{M}(R)/R^2$, where $\mathcal{M}(R)$ is the mass interior to R. Within the bulge $v_\theta(R)$ is observed to rise steeply with R, as would be expected for any reasonable density profile, for which $\mathcal{M}(R)$ increases rapidly with R. The contribution of the bulge to $\mathcal{M}(R)$ is of course constant outside the bulge, so for a spherically symmetric bulge as long as the bulge dominates the gravitational potential $v_\theta(R)$ is expected to approach the Keplerian result $v_\theta(R) \propto R^{-1/2}$, as is observed.

But this decrease in v_θ does not continue. The contribution of the disk to v_θ rises more or less linearly as R increases from zero to about 5 kpc. The disk begins to contribute appreciably for $R > 1$ kpc, and becomes dominant at $R \simeq 3$ kpc, where v_θ is observed to reach a minimum, and then to begin to rise. The calculation of the gravitational potential due to a thin disk is more complicated than for the case of spherical symmetry, and is briefly described in Appendix A of this section. As noted there, for the special case of rigid rotation for which $v_\theta(R) \propto R$, the mass per area $\Sigma(R)$ of the disk must also grow as R. As shown in Appendix B of this section, this rigid rotation is to be expected for

---

[14] For instance, for distances from the center varying from less than 1 kpc to 20 kpc, see Y. Sofue, M. Honma, and T. Omadaka, *Publ. Astron. Soc. Japan* **61**, 227 (2009). For later results covering distances from 4 kpc to 22 kpc, see D. Russell *et al.*, *Astron. Astrophys.* **601**, L5 (2017). Other references are given in a review article by J. Bland-Harwood and O. Gerhard, *Ann. Rev. Astron. Astrophys.* **59**, 529 (2016).

a system that has settled into a state of lowest energy while keeping a constant angular momentum.

The rise with R of $v_\theta(R)$ does not persist. For R greater than about 5 kpc v_θ is roughly constant, as first noted for the Andromeda galaxy M31 by Rubin and Ford.[15] In particular, this is true of the solar neighborhood, for which $R \simeq 8$ kpc. As shown in Appendix 4.3A, as long as the gravitational potential is dominated by the disk, this requires that the surface mass density Σ of the disk falls off as $1/R$. Also, this leveling of $v_\theta(R)$ shows that the outer part of the galaxy has not settled into a state of lowest energy.

It is now believed that for R greater than about 20 kpc the gravitational potential is dominated by a spherical halo of dark matter rather than the disk. For a spherically symmetric distribution of mass, the velocity $v_\theta(r)$ in circular orbits of radius r is again related to the mass $\mathcal{M}(r)$ within a radius r by $v_\theta^2(r)/r = G\mathcal{M}(r)/r^2$, so in a range of radii where $v_\theta(r)$ is constant, we must have $\mathcal{M}(r) \propto r$. In 1974 Ostriker, Peebles, and Yahil[16] noted that if most of the mass of disk galaxies is supposed to occupy a spherical halo, extending far beyond the disk, then observed velocities within local giant spiral galaxies indicate that the mass $\mathcal{M}(r)$ within a radius r does increase linearly with r, for r between 20 kpc and 500 kpc. As they noted, and as remarked in the previous section, this is what would be expected if the halo were an isothermal sphere. Similar conclusions were reached at about the same time by Einasto, Kaasik, and Saar.[17] As pointed out by these two groups, the great increase in the estimated mass of galaxies alleviated the problem that previously the estimated mass in the galaxies in clusters and in intracluster gas was not sufficient to satisfy the requirements of the virial theorem of Section 1.1 for the stability of clusters. Since the newly recognized mass was not observed optically, it came to be known as *dark matter*. (Another factor that seemed important at the time was that, with only galactic masses taken into account in calculating the rate of cosmic expansion, the previous estimates of galactic masses had given a cosmic mass density much less than indicated by measurements of the Hubble constant under the assumption of zero spatial curvature. In fact, although the increase in estimated galactic mass reduced this discrepancy, most of the energy governing the expansion of the universe is a vacuum energy discovered later, in 1998.[18])

The flattening of the curve of velocity versus radius did not in itself require that this mass take the form of a spherical halo. There was still the possibility

---

[15] V. C. Rubin and W. K. Ford, *Astrophys. J.* **159**, 379 (1970). For further work on the rotation curves of various galaxies by Rubin and her collaborators, see V. Rubin, N. Thonnard, and W. K. Ford, Jr., *Astrophys. J.* **238**, 471 (1980); V. Rubin, D. Burstein, W. K. Ford, Jr., and N. Thonnard, *Astrophys. J.* **289**, 81 (1985); V. Rubin, J. A. Graham, and J. D. P. Kenney, *Astrophys. J.* **394**, L9 (1992).
[16] J. P. Ostriker, P. J. E. Peebles, and A. Yahil, *Astrophys. J.* **193**, L1 (1974).
[17] J. Einasto, A. Kaasik, and E. Saar, *Nature* **250**, 309 (1974).
[18] S. Perlmutter *et al.* (Supernova Cosmology Project), *Astrophys. J.* **517**, 565 (1999); A. G. Riess *et al.* (High z-Supernova Search Team), *Astron. J.* **116**, 1099 (1998).

that much of the dark matter inhabited the plane of the galaxy. To get constant orbital velocities, we need a gravitational force per mass that goes as $1/R$. (We again use an upper case R here and below to distinguish the radial coordinate of cylindrical coordinates from the radial coordinate r of spherical coordinates.) As shown in Appendix A of this section, this force would be produced by a mass per area that decreases as $1/R$, giving a mass interior to the radius R that goes as R, just the same as found for a spherical halo. Ostriker and Peebles argued for a halo, as avoiding the instability[19] of a cold disk against non-axisymmetric perturbations.

Presumably baryonic matter in spiral galaxies has fallen into the galactic plane, losing energy by radiation, while preserving its angular momentum. It is more difficult for the dark matter of the halo to lose energy, because it cannot radiate, but some of it is dragged from the halo into the disk by the gravitational field of the baryonic matter.[20]

Appendix A: The Gravitational Potential of a Disk

This appendix will calculate the gravitational potential due to a thin disk.[21] It is believed that this is the dominant contribution to the gravitational potential of disk galaxies in an intermediate range of distances from the center. As we will see, this calculation of the gravitational potential is more complicated than for a spherically symmetric mass distribution.

We use cylindrical coordinates R, θ, z, with the disk in the plane $z = 0$, so that its mass density is

$$\rho(R, \theta, z) = \delta(z) \Sigma(R, \theta), \tag{4.3.A1}$$

with Σ the mass per area on the disk. We are here allowing an arbitrary dependence on the angular coordinate θ, because this departure from cylindrical symmetry turns out to cause little extra difficulty in the calculation.

In cylindrical coordinates, the Poisson equation $\nabla^2 \phi = 4\pi G \rho$ reads

$$\left[\frac{1}{R} \frac{\partial}{\partial R} \left(R \frac{\partial}{\partial R} \right) + \frac{1}{R^2} \frac{\partial^2}{\partial \theta^2} + \frac{\partial^2}{\partial z^2} \right] \phi(R, \theta, z) = 4\pi G \delta(z) \Sigma(R, \theta). \tag{4.3.A2}$$

First let us consider solutions of the Poisson equation for $z \neq 0$, in which case it is just the Laplace equation $\nabla^2 \phi = 0$. There is a well-known class of solutions that are finite at $z = 0$ and vanish as $|z| \to \infty$:

---

[19] J. P. Ostriker and P. J. E. Peebles, *Astrophys. J.* **186**, 467 (1973).
[20] G. R. Flores, S. M. Faber, R. Flores, and J. R. Primack, *Astrophys. J.* **301**, 27 (1986).
[21] The discussion in this appendix is based on the treatment of J. Binney and S. Tremaine, *Galactic Dynamics* (Princeton University Press, Princeton, NJ, 1987).

$$\phi_{\nu,k}(R,\theta,z) = J_{|\nu|}(kR)e^{i\nu\theta}e^{-k|z|}, \tag{4.3.A3}$$

where ν is an arbitrary integer; k is an arbitrary positive real number; and J_ν is the Bessel function of order ν, satisfying the differential equation

$$\frac{1}{x}\frac{d}{dx}\left(x\frac{dJ_\nu(x)}{dx}\right) + \left(1 - \frac{\nu^2}{x^2}\right)J_\nu(x) = 0. \tag{4.3.A4}$$

These solutions are finite and continuous at the disk, where $z = 0$, but their first derivatives with respect to z are discontinuous. As $z \to 0$ from above or below, we have

$$\left.\frac{\partial \phi_{\nu,k}(R,\theta,z)}{\partial z}\right|_{z\to 0\pm} = \mp k J_{|\nu|}(kR)e^{i\nu\theta},$$

so

$$\left.\frac{\partial^2 \phi_{\nu,k}(R,\theta,z)}{\partial z^2}\right|_{z\to 0} = -2k\delta(z)J_{|\nu|}(kR)e^{i\nu\theta}.$$

Comparing this with Eq. (4.3.A2), we see that the potential $\phi_{\nu,k}(R,\theta,z)$ is produced by a term $-2kJ_{|\nu|}(kR)e^{i\nu\theta}/4\pi G$ in the surface mass density. Hence if we write a general surface density distribution as a superposition

$$\Sigma(R,\theta) = \mathrm{Re}\sum_{\nu=0}^{\infty}\int_0^\infty dk\, \mu_{\nu,k} k J_\nu(kR)e^{i\nu\theta}, \tag{4.3.A5}$$

then the solution of Eq. (4.3.A2) is

$$\phi(R,\theta,z) = -2\pi G\,\mathrm{Re}\sum_{\nu=0}^{\infty}\int_0^\infty dk\, \mu_{\nu,k} J_\nu(kR)e^{i\nu\theta}e^{-k|z|}, \tag{4.3.A6}$$

and the radial gravitational force per test-body mass is

$$\frac{\partial}{\partial R}\phi(R,\theta,z) = -2\pi G\,\mathrm{Re}\sum_{\nu=0}^{\infty}\int_0^\infty dk\, \mu_{\nu,k} k J'_\nu(kR)e^{i\nu\theta}e^{-k|z|}. \tag{4.3.A7}$$

By the way, quite general functions $\Sigma(R,\theta)$ of R and θ can be expressed as a superposition of the form (4.3.A5), known as Hankel transforms. Under very general conditions we can write Σ as a Fourier series

$$\Sigma(R,\theta) = \mathrm{Re}\sum_{\nu=0}^{\infty} e^{i\nu\theta}\Sigma_\nu(R).$$

To write $\Sigma_\nu(R)$ as a superposition of Bessel functions, we note that for any positive-definite a and b,

$$\int_0^\infty ak\,dk\, J_\nu(ka)J_\nu(kb) = \delta(a-b).$$

Therefore, by taking

$$\mu_{\nu,k} = \int_0^\infty dR\, R J_\nu(kR) \Sigma_\nu(R),$$

we can write

$$\Sigma_\nu(R) = \int_0^\infty dk\, \mu_{\nu,k} k J_\nu(kR),$$

which gives the desired expression Eq. (4.3.A5).

As a special case of some importance, note that a surface mass distribution with a θ-independent $\Sigma(R,\theta) \propto R^n$ is obtained if $\mu_{\nu,k} \propto k^{-n-2}\delta_{\nu,0}$, so, where the gravitational potential is dominated by the mass of the disk, the gravitational force per mass (4.3.A7) goes as R^n. Setting this equal to v_θ^2/R gives $v_\theta \propto R^{(n+1)/2}$. In particular, to get a rigid rotation, with $v_\theta \propto R$, we must have $n = 1$, in which case $\Sigma(R,\theta) \propto R$. In this case the Hankel representation (4.3.A5) does not converge for $k \to 0$. On the other hand, as long as the gravitational potential is dominated by the matter of the disk, to get the R-independent circular velocity observed in the outer parts of the disk we would need $n = -1$, in which case Σ would be proportional to $1/R$. (It should be noted that for $n = -1$ and any ν the integrals in Eqs. (4.3.A5) and (4.3.A7) for the surface mass density and the gravitational force both converge. For $k \to 0$ we have $J_\nu(kR) \to (kR)^\nu/\nu!$, so the integrals converge at $k = 0$, while for $k \to \infty$ we have $J_\nu(kR) \to (2/\pi kR)^{1/2}\cos(kR - \nu\pi/2 - \pi/4)$, so the oscillation of the Bessel functions makes the integrals converge at $k \to \infty$.)

There is an important qualitative difference between the gravitational potentials produced by distributions of mass that are spherically symmetric in three dimensions or circularly symmetric in two dimensions. As first proved by Newton, in the case of spherical symmetry the gravitational acceleration $\phi'(r)$ at a distance r from the center of symmetry is $-G\mathcal{M}(r)/r^2$, and so depends on no aspect of the mass distribution other than the mass $\mathcal{M}(r) = 4\pi \int_0^r \rho(r')r'^2\, dr'$ interior to the radius r. This is not true in two dimensions even for a circularly symmetric surface distribution of mass. For instance, consider the surface mass distribution $\Sigma(R) = C(a^2 + R^2)^{-3/2}$, with arbitrary constants C and a. This is given[22] by Eq. (4.3.A5) if we take $\mu_{0,k} = (C/a)\exp(-ak)$ and $\mu_{\nu,k} = 0$ for $\nu \neq 0$. In this case, Eq. (4.3.A6) gives the gravitational potential

$$\phi(R) = -(2\pi GC/a)(a^2 + R^2)^{-1/2} = -G[\mathcal{M}(\infty) - \mathcal{M}(R)]/a,$$

which depends not only on the function $\mathcal{M}$, but also on the length scale a.

---

[22] The integrals needed in this discussion are given in §7 of W. Magnus and F. Oberhettinger, *Formulas and Theorems for the Functions of Mathematical Physics* (Chelsea Publishing Co., New York, 1949).

Appendix B: Minimum Energy for Fixed Angular Momentum

Consider a distribution of mass with density $\rho(\mathbf{x})$ and mean velocity $\mathbf{v}(\mathbf{x})$.[23] The angular momentum vector is

$$\mathbf{J} = \int d^3x \, \rho(\mathbf{x}) \mathbf{x} \times \mathbf{v}(\mathbf{x}). \tag{4.3.B1}$$

We consider variations in $\mathbf{v}(\mathbf{x})$ with $\mathbf{J}$ fixed, and for the moment with $\rho(\mathbf{x})$ not varied. Since we are not varying $\rho(\mathbf{x})$, the gravitational energy is not varied. It therefore suffices for us to consider only the effect of the variation in $\mathbf{v}(\mathbf{x})$ on the kinetic energy

$$T = \frac{1}{2} \int d^3x \, \rho(\mathbf{x}) \mathbf{v}^2(\mathbf{x}). \tag{4.3.B2}$$

We will use cylindrical coordinates R, θ, z, with the z-axis chosen to be in the direction of $\mathbf{J}$, so that the magnitude of the angular momentum is

$$J = \int d^3x \, \rho(\mathbf{x}) R \, v_\theta(\mathbf{x}). \tag{4.3.B3}$$

Evidently

$$T \geq \frac{1}{2} \int d^3x \, \rho(\mathbf{x}) v_\theta^2(\mathbf{x}), \tag{4.3.B4}$$

with equality only if the orbits are circles in the plane with fixed z. Hence to minimize T we must take $v_R = v_z = 0$, which does not require any change in J. Now, by the Schwarz inequality

$$\int d^3x \, \rho(\mathbf{x}) v_\theta^2(\mathbf{x}) \times \int d^3x \, \rho(\mathbf{x}) R^2 \geq \left[\int d^3x \, \rho(\mathbf{x}) v_\theta(\mathbf{x}) R \right]^2 = J^2, \tag{4.3.B5}$$

with equality only for $v_\theta(\mathbf{x}) \propto R$, that is, for rigid rotation. Adjusting $v_\theta(R)$ to be proportional to R can be done without any change in J, as we can take $v_\theta = RJ/I$, where $I = \int d^3x \, \rho(\mathbf{x}) R^2$. Hence, for a fixed $\rho(\mathbf{x})$ and J, the energy is minimized by taking the velocity to be a rigid rotation around the angular momentum vector, with fixed angular velocity $v_\theta/R = J/I$. The total energy may be further reduced by adjustment of the mass density $\rho(\mathbf{x})$, but whatever density is chosen, the energy can always then be further minimized by letting the velocity be a rigid rotation around the angular momentum vector. Since the velocities of stars in the outer parts of galactic disks are not proportional to their radial coordinate, but approximately constant, we can conclude that these disks have not relaxed by angular momentum conserving processes to a state of minimum energy.

---

[23] The discussion here is based on that of D. Lynden-Bell and J. E. Pringle, *Mon. Not. Roy. Astron. Soc.* **168**, 603 (1974).

4.4 Spiral Arms

We now focus on the spiral arms of the disk galaxies we have been considering. As pointed out in a recent article[24] on this subject, "Seventy percent of galaxies in the nearby universe are characterized by a disk with prominent spiral arms, but our understanding of the origin of these patterns is incomplete, even after decades of theoretical study." Apparently spiral arms are regions of increased gas density, because they are marked with relatively short-lived massive bright stars, which must be continually forming. But what produces these spiral inhomogeneities?

One thing seems clear: unless dissipative effects such as viscosity and heat conduction play a significant role, which is not usually assumed to be the case, spiral arms are not equilibrium solutions of the equations of galactic dynamics. In the absence of dissipation, these equations are invariant under time-reversal, so any solution with a given time-dependence must be accompanied by another with the opposite time-dependence. In particular, a solution with trailing spiral arms, in which the distance R of co-moving points on each arm from the symmetry axis decreases along the arm in the direction of galaxy rotation, must be accompanied by another solution with leading spiral arms, in which R increases in this direction. But very few if any galaxies have leading spiral arms.

Note that the absence of leading arms contradicts time-reversal symmetry only if these are equilibrium solutions — there is no contradiction if the solutions with trailing spiral arms are the ones that grow with time, while the ones with leading arms decay. We will see an example of this below.

A further argument against spiral arms being an equilibrium configuration is provided by the point discussed in Appendix B of the previous section: A disk of fixed angular momentum that has relaxed to a configuration of minimum energy would have to be rotating rigidly, with fixed angular velocity, in conflict with the dependence on R observed for angular velocities outside a central region.

The idea that spiral arms represent a wave of enhanced density passing through a disk galaxy was inspired in part by a 1964 paper of Lin and Shu.[25] They considered matter moving in a plane, with negligible velocity dispersion. In cylindrical coordinates the equations governing the surface density $\Sigma(R,\theta,t)$ and velocity components $v_R(R,\theta,t)$ and $v_\theta(R,\theta,t)$ are the equation of continuity

$$\frac{\partial}{\partial t}\Sigma + \frac{1}{R}\frac{\partial}{\partial R}(R\Sigma v_R) + \frac{\partial}{\partial \theta}(\Sigma v_\theta) = 0, \qquad (4.4.1)$$

---

[24] E. D'Onghia, M. Vogelsberger, and L. Hernquist, *Astrophys. J.* **766**, no. 1, paper 34 (2013).
[25] C. C. Lin and F. H. Shu, *Astrophys. J.* **140**, 646 (1964).

and the R and θ components of the Euler equation for zero viscosity:

$$\frac{\partial}{\partial t}v_R + v_R\frac{\partial v_R}{\partial R} + \left(\frac{v_\theta}{R}\right)\frac{\partial v_R}{\partial \theta} - \frac{v_\theta^2}{R} = -\frac{\partial \phi}{\partial R}, \qquad (4.4.2)$$

$$\frac{\partial}{\partial t}v_\theta + v_R\frac{\partial v_\theta}{\partial R} + \left(\frac{v_\theta}{R}\right)\frac{\partial v_\theta}{\partial \theta} + \frac{v_\theta v_R}{R} = -\frac{1}{R}\frac{\partial \phi}{\partial \theta}. \qquad (4.4.3)$$

There is also the Poisson equation for $\phi(R,\theta,z)$. All these equations are evidently invariant under the time-reversal transformation, which reverses the sign of the time and all velocity components.

These equations have a familiar unperturbed solution, distinguished by a subscript 0, with $v_{0R} = 0$ and $v_{0\theta}$, Σ_0, and ϕ_0 functions only of R, governed by the R component of the Euler equation

$$v_{0\theta}^2(R)/R = \frac{\partial \phi_0(R)}{\partial R}, \qquad (4.4.4)$$

and the unperturbed Poisson equation. We consider an infinitesimal perturbation, distinguished by a prefix δ, so that $\Sigma = \Sigma_0 + \delta\Sigma$, and likewise for v_R, v_θ, and ϕ. To first order in perturbations, Eqs. (4.4.1)–(4.4.3) become

$$\frac{\partial}{\partial t}\delta\Sigma + \frac{1}{R}\frac{\partial}{\partial R}(R\Sigma_0\,\delta v_R) + \Sigma_0\frac{\partial}{\partial \theta}(\delta v_\theta) + v_{0\theta}\frac{\partial}{\partial \theta}(\delta\Sigma) = 0, \qquad (4.4.5)$$

$$\frac{\partial}{\partial t}\delta v_R + \left(\frac{v_{0\theta}}{R}\right)\frac{\partial \delta v_R}{\partial \theta} - 2\frac{v_{0\theta}\,\delta v_\theta}{R} = -\frac{\partial \delta\phi}{\partial R}, \qquad (4.4.6)$$

$$\frac{\partial}{\partial t}\delta v_\theta + \delta v_R\frac{\partial v_{0\theta}}{\partial R} + \left(\frac{v_{0\theta}}{R}\right)\frac{\partial \delta v_\theta}{\partial \theta} + \frac{v_{0\theta}\,\delta v_R}{R} = -\frac{1}{R}\frac{\partial \delta\phi}{\partial \theta}. \qquad (4.4.7)$$

Just for illustration, in the unlikely case that the galaxy's mass is entirely contained in its disk, the Poisson equation (4.3.A2) becomes

$$\left[\frac{1}{R}\frac{\partial}{\partial R}\left(R\frac{\partial}{\partial R}\right) + \frac{1}{R^2}\frac{\partial^2}{\partial \theta^2} + \frac{\partial^2}{\partial z^2}\right]\delta\phi(R,\theta,z) = 4\pi G\delta(z)\,\delta\Sigma(R,\theta). \qquad (4.4.8)$$

Because the unperturbed quantities are independent of time and θ, any solution of Eqs. (4.4.5)–(4.4.7) and the perturbed Poisson equation remains a solution if we translate θ and t by any constant amounts. Barring degeneracies, this translation must therefore give the same solution up to a constant factor, so we expect to find solutions with the angular and time-dependence

$$\delta\Sigma(R,\theta,t) = e^{-i\omega t + i\nu\theta}\overline{\Sigma}(R), \qquad (4.4.9)$$

$$\delta v_R(R,\theta,t) = e^{-i\omega t + i\nu\theta}\overline{v_R}(R), \qquad (4.4.10)$$

$$\delta v_\theta(R,\theta,t) = e^{-i\omega t + i\nu\theta}\overline{v_\theta}(R), \qquad (4.4.11)$$

$$\delta\phi(R,\theta,z,t) = e^{-i\omega t + i\nu\theta}\overline{\phi}(R,z). \qquad (4.4.12)$$

These functions must not change when θ is shifted by a multiple of 2π, so ν must be a real positive or negative integer, while ω can be any complex number.

Of course, the physical perturbation will have components that are the real parts of superpositions of (4.4.9)–(4.4.12). We will consider just a single value of ν and of ω, assuming that these dominate the superposition.

Now, we can take $\overline{\Sigma}(R) = S(R)\exp(i\Phi(R))$, where $S(R)$ and $\Phi(R)$ are real. In this case, the density perturbation is

$$\text{Re}\big(\delta\Sigma(R,\theta,t)\big) = S(R)\exp\big(\text{Im}(\omega)\,t\big)\cos\big(\Phi(R)-\text{Re}(\omega)\,t+\nu\theta\big). \quad (4.4.13)$$

If the phase $\Phi(R)$ varies with R much more rapidly than the modulus $S(R)$, then the maxima of the density at time t will be at values of R and θ for which the cosine is a maximum, that is, for

$$\Phi(R) - \text{Re}(\omega)\,t + \nu\theta = 2\pi N, \quad (4.4.14)$$

where N is an integer. This represents only $|\nu|$ independent curves, because shifting N by $|\nu|$ gives a solution with θ shifted by $2\pi|\nu|/\nu = \pm 2\pi$, which of course has no effect on the curve.

We can choose to look at the galaxy from the side where it appears to rotate counter-clockwise – that is, with θ increasing with time. Then ν and $\text{Re}\,\omega$ have the same sign. Since the cosine in Eq. (4.4.13) is an even function of its argument, we can choose this common sign to be positive. Then if $\Phi(R)$ increases or decreases monotonically, these $|\nu|$ curves are respectively trailing or leading spirals, for which at fixed time R steadily decreases or increases with increasing θ, the direction of galaxy rotation.

By the way, time-reversal invariance tells us that for each growing mode with $\text{Im}\,\omega > 0$ there is a decaying mode with $\text{Im}\,\omega < 0$, Since also the sign of $\text{Re}\,\omega$ is reversed, in this time-reversed solution the galaxy is rotating clockwise, toward decreasing θ, so that trailing spirals become leading spirals, and vice versa.

With $\Phi(R)$ varying much more rapidly than other quantities, it is even possible to solve Eqs. (4.4.5)–(4.4.8) using the WKB approximation of wave mechanics. But these happy results depend on a big "If." The usual WKB approximation of wave mechanics depends on the assumption that the coefficients in the differential equation should change little in a distance equal to the effective wave length, which here is $2\pi/|\Phi'(R)|$. It is possible to focus on special cases where this assumption is true, such as tightly wound spirals, but there is no reason to expect it to be the case for general spiral galaxies, because the equations (4.4.5)–(4.4.8) involve no small dimensionless parameters. In any case, it is necessary to go beyond the approximation of infinitesimal perturbations to see how the exponential growth can cease and the spiral arms yet survive. Apparently Lin lost his enthusiasm for this particular model of density waves, because it is barely mentioned in a 1996 book by Bertin and Lin.[26]

---

[26] G. Bertin and C. C. Lin, *Spiral Structure in Galaxies: A Density Wave Theory* (MIT Press, Cambridge, MA, 1996).

The differential rotation of the disk at an angular frequency $\Omega(R)$ that depends on R did not play any explicit role in the appearance of spiral structure in Eq. (4.4.13), though of course it must affect whatever solutions are found for $S(R)$ and $\Phi(R)$. To see directly how differential rotation can lead to spiral structures, it is instructive to consider a generalization of a "thought experiment" described by Binney and Tremaine.[27]

Consider some inhomogeneity that initially at $t = 0$ occupies a curve, described by some formula $\theta = \Theta_0(R)$ for the azimuthal angle at a given radius. (Binney and Tremaine took this curve to be a straight line extending in a radial direction, but almost any function Θ_0 will do, as long as R extends over a sufficient range of radii.) If the velocities are circular but non-rigid, with components that are only in the θ-direction and equal to a function $v_\theta(R)$ of the radial coordinate R, then at a later time t the inhomogeneity will lie on the curve

$$\theta = \Theta_0(R) + \Omega(R)t, \qquad (4.4.15)$$

where $\Omega(R) \equiv v_\theta(R)/R$. If $\Omega(R)$ has the same sign for all R, and increases or decreases monotonically with R, then after a sufficient time has passed Eq. (4.4.15) will be the equation of a spiral. For instance, if $\Omega(R) > 0$ and $\Omega'(R) < 0$ for all R, then at late times any point on the curve (4.4.15) will move in the direction of increasing θ, and (since $dR/d\theta = 1/\Omega'(R) < 0$) the spiral will curl inward with increasing θ. In this case, we have a trailing spiral. More generally, at late times the spiral is trailing if $\Omega'/\Omega < 0$, and leading if $\Omega'/\Omega > 0$. Since $\Omega(R) \propto 1/R$ for the outer parts of typical spiral galaxies, we have $\Omega'/\Omega \simeq -1/R$, so the spiral arm is trailing, in agreement with the observation that galaxy spiral arms are usually trailing.

There is a problem with this, that is often expressed in terms of the pitch angle. This is the angle α between the θ-direction at some point on the spiral and the tangent to the spiral at that point. We can draw an infinitesimal right triangle, with one side in the θ-direction with length equal to some infinitesimal displacement $R(\theta)\,\delta\theta$, and another side in the perpendicular R-direction with length equal to the magnitude $|R'(\theta)|\,\delta\theta$ of the change in R when θ is changed by $\delta\theta$. The hypotenuse then lies along the direction of the spiral, and the pitch angle α is the angle between the hypotenuse and the side in the θ-direction. Its cotangent is the ratio of the adjacent to the opposite side, so in general

$$\cot\alpha = \frac{R(\theta)\,\delta\theta}{|R'(\theta)|\,\delta\theta} = R\left|\frac{d\theta(R)}{dR}\right|. \qquad (4.4.16)$$

For the particular spiral (4.4.15), at late times this is $\cot\alpha = Rt|\Omega'(R)|$. If $\Omega(R) \propto 1/R$, we have $\cot\alpha = t|\Omega(R)|$, which is the angle traced out by a

---

[27] J. Binney and S. Tremaine, *Galactic Dynamics*, 2nd edn. (Princeton University Press, Princeton, NJ, 2008).

co-moving star at radial coordinate R in a time t. But for reasonable parameters, this is very large, because co-moving matter in the neighborhood of spiral arms in the galaxies we observe has typically already made many turns around the galactic center, driving α to a very small value. For instance, for the parameters taken as a typical example by Binney and Tremaine, $R = 5$ kpc, $\Omega(R) = 200$ km/sec/R [kpc], and $t = 10^{10}$ years, the pitch angle turns out to be $0.14°$. We do not see pitch angles this small in spiral galaxies, which typically have pitch angles of order $10°$. This is called the *winding problem*. It would be avoided if we supposed that the ages of the spiral arms we see are no more than about 10^8 years, rather than 10^{10} years, but this seems unlikely.

We now turn to a different approach, which relies on a closer analysis of the actual motion of stars in a disk galaxy and though incomplete seems to have become a consensus approach to the problem of spiral arms. As we will see, the winding problem is still present in this approach, but is much milder.

The familiar equations of planar motion of a star in a gravitational potential $\phi(R)$ are

$$\ddot{R} = R\dot{\theta}^2 - \phi'(R) \tag{4.4.17}$$

and

$$\frac{d}{dt}(R^2\dot{\theta}) = 0. \tag{4.4.18}$$

As a consequence of Eq. (4.4.18), we can define a time-independent quantity

$$L = R^2\dot{\theta} \tag{4.4.19}$$

and write Eq. (4.4.17) as

$$\ddot{R} = \frac{L^2}{R^3} - \phi'(R). \tag{4.4.20}$$

For a circular orbit with time-independent radius R_0, $\dot{\theta}$ takes the constant value $\Omega(R_0)$, where according to Eq. (4.4.20)

$$\Omega^2(R_0) = L^2(R_0)/R_0^4 = \phi'(R_0)/R_0. \tag{4.4.21}$$

For nearly circular orbits, we have

$$R(t) = R_0 + \Delta(t) \tag{4.4.22}$$

with $|\Delta| \ll R_0$. Equation (4.4.20) then becomes

$$\ddot{\Delta} + \kappa^2(R_0)\Delta = 0, \tag{4.4.23}$$

where κ is the *epicyclic frequency*

$$\kappa^2(R_0) = \frac{3L^2(R_0)}{R_0^4} + \phi''(R_0) = 3\phi'(R_0)/R_0 + \phi''(R_0). \tag{4.4.24}$$

Thus the orbit exhibits two kinds of periodic motion: θ increases (or decreases) at an average rate $\Omega(R_0)$, while R oscillates around R_0 at a rate $\kappa(R_0)$, which according to Eqs. (4.4.24) and (4.4.21) can be expressed in terms of Ω and its derivative

$$\kappa^2(R_0) = 4\Omega^2(R_0) + 2\Omega(R_0)\Omega'(R_0)R_0. \tag{4.4.25}$$

The orbit cannot close unless Ω and κ are *commensurate* – that is, unless κ/Ω is a ratio p/q of whole numbers p and q. In that case, R executes p whole oscillations in a time when θ increases by $2\pi q$. For instance, for a potential $\phi \propto 1/R$, $\kappa(R_0) = \pm\Omega(R_0)$, whereas for a potential $\phi(R) \propto R^2$, $\kappa = \pm 2\Omega$, and in both cases the orbits are closed curves. But this is not generally the case.

Lindblad[28] pointed out that in our galaxy $\Omega(R_0) - \kappa(R_0)/2$ is nearly independent of R_0. Hence we can imagine a perturbation to circular orbits, for which orbits close in a frame of reference that rotates at a smaller "pattern frequency"

$$\Omega_p = \Omega(R_0) - \kappa(R_0)/2 \tag{4.4.26}$$

(with any convenient value of R_0). In this frame a star at R_0 will revolve at an angular frequency $\Omega(R_0) - \Omega_p = \kappa(R_0)/2$, so the orbits of stars will appear as closed slightly elliptical time-independent curves, on which R goes through two complete cycles of increase and decrease in the time $4\pi/\kappa(R_0)$ in which the star makes one circuit of its orbit. Also, $\kappa(R_0)$ is near $2\Omega(R_0)$ over a sizable range of radii, so the pattern frequency Ω_p is considerably smaller than $\Omega(R_0)$.

It was noted by Kalnajs[29] that when stars are spread evenly on a family of these closed curves, with the axes of the curves gradually changing their orientation from one curve to another, the stars tend to crowd together at certain spots on their orbits. To identify these crowded spots, note that the radial coordinate of the star on an orbit of mean radius R_0, at a time t after passing a point of maximum distance from the center of the orbit, is

$$R(t) = R_0\left(1 + e\cos\bigl(\kappa(R_0)t\bigr)\right), \tag{4.4.27}$$

where e is the positive parameter known as the eccentricity, and is usually assumed to be considerably less than unity. (This holds both in an inertial and in a rotating frame.) Also, if the point on this orbit at maximum distance $R_0(1+e)$ from the center lies in a direction $\Theta(R_0)$, then at the same time t, the angular coordinate of the star in the frame rotating with pattern speed Ω_p is

$$\theta(t) = \Theta(R_0) + \bigl(\Omega(R_0) - \Omega_p\bigr)t = \Theta(R_0) + \kappa(R_0)t/2, \tag{4.4.28}$$

so Eq. (4.4.27) can be written (dropping the time argument)

$$R = R_0\left(1 + e\cos\bigl(2[\theta - \Theta(R_0)]\bigr)\right). \tag{4.4.29}$$

---

[28] B. Lindblad, *Stockholms Ann.* **19**, No. 7 (1956); **20**, No. 4 (1958). His work on spiral arms goes back to B. Lindblad, *Stockholms Ann.* **12**, No. 4 (1936). This is the astronomer Bertil Lindblad, not to be confused with the physicist Goran Lindblad, who derived the equation that governs the evolution of the density matrix in the quantum theory of open systems.

[29] A. J. Kalnajs, *Proc. Astron. Soc. Australia* **2**, 174 (1973).

(There is a correction of order e to Eq. (4.4.28) due to variations in the velocity of the star as its radial coordinate increases and decreases, but taking these into account would lead to corrections of order e^2 in Eq. (4.4.29), and it can therefore be neglected.) Now, if we have two orbits with nearly equal mean radii R_0 and $R_0 + \delta R_0$, then the distance between stars on these orbits when both are at direction θ is

$$(R_0 + \delta R_0)\left(1 + e\cos\left(2[\theta - \Theta(R_0 + \delta R_0)]\right)\right)$$
$$- R_0\left(1 + e\cos\left(2[\theta - \Theta(R_0)]\right)\right)$$
$$= \delta R_0 \left[1 + e\cos\left(2[\theta - \Theta(R_0)]\right) - 2eR_0\Theta'(R_0)\sin\left(2[\theta - \Theta(R_0)]\right)\right]$$
$$+ O(\delta R_0^2). \tag{4.4.30}$$

This has a minimum at an angle $\theta_m(R_0)$ at which its first derivative with respect to θ vanishes and its second derivative is positive:

$$\sin\left(2[\theta_m(R_0) - \Theta(R_0)]\right) + 2R_0\Theta'(R_0)\cos\left(2[\theta_m(R_0) - \Theta(R_0)]\right) = 0, \tag{4.4.31}$$

$$-\cos\left(2[\theta_m(R_0) - \Theta(R_0)]\right) + 2R_0\Theta'(R_0)\sin\left(2[\theta_m(R_0) - \Theta(R_0)]\right) > 0. \tag{4.4.32}$$

For instance, if (as suggested by Kalnajs) the dimensionless parameter $R_0\Theta'(R_0)$ is large and negative, then these conditions are satisfied by

$$\theta_m(R_0) \simeq \Theta(R_0) + 3\pi/4 \quad \text{and} \quad \theta_m(R_0) \simeq \Theta(R_0) + 7\pi/4,$$

whereas if $R_0\Theta'(R_0)$ is large and positive, then

$$\theta_m(R_0) \simeq \Theta(R_0) + \pi/4 \quad \text{and} \quad \theta_m(R_0) \simeq \Theta(R_0) + 5\pi/4.$$

In any case, the crowded spots fall on two curves of θ_m vs. R_0, which are spirals if neither $\theta_m(R_0)$ nor $\theta'_m(R_0)$ changes sign over a wide range of R_0. Assuming that the galaxy revolves in the direction of increasing θ, they are trailing spirals if $\theta'_m(R_0)/\theta_m(R_0) < 0$ and leading spirals if $\theta'_m(R_0)/\theta_m(R_0) > 0$. Note that here there are *two* of these spirals, This agrees with the observation that most spiral galaxies, such as the beautiful spirals M100, M81, and M51 (the galaxy in which spiral arms were first observed, by the Earl of Rosse in 1850), have two distinct arms.

It is not essential to adopt this picture of the origin of a curve of enhanced density, or even that this curve is a spiral. The expression $\Omega_p = \Omega(R) - \kappa(R)/2$ for the pattern speed in Eq. (4.4.26) actually has a non-negligible though weak dependence on radius. Thus, if in an inertial frame at $t = 0$ the points where stars are most crowded lie on a curve $\theta = \theta_m(R)$, then at a later time t they will lie on the curve

$$\theta = \theta_m(R) + \Omega_p(R)t = \theta_m(R) + [\Omega(R) - \kappa(R)/2]t. \tag{4.4.33}$$

Just as before, whatever the shape of this curve, as long as $\Omega(R) - \kappa(R)/2$ and its first derivative with respect to R does not change sign over a range of R, the curve will eventually wind up to a spiral.

There is also a potential winding problem here. Repeating the above discussion that led to the result that at late time the pitch angle α is given by $\cot\alpha \to R|\Omega'(R)|$, we now have a pitch angle α at late time with

$$\cot\alpha \to Rt\left|\frac{d\Omega_p(R)}{dR}\right| = Rt\left|\frac{d[\Omega(R) - \kappa(R)/2]}{dR}\right|. \qquad (4.4.34)$$

According to Binney and Tremaine, in our galaxy for R between 5 and 10 kpc, $R\,d\Omega_p(R)/dR$ has a mean value of about 7 km/sec per kiloparsec, so for $t = 10^{10}$ years the pitch angle would be $\alpha \simeq 0.8°$, considerably larger than before. The winding problem here is avoided if spiral arms are no more than 10^9 years old, but this is still unlikely.

So far, we have considered only the orbits of stars in a fixed galactic gravitational field. The rotating pattern we have described produces a perturbation to this gravitational field; we need to consider whether the perturbed field tends to preserve the pattern, or to destroy it.

At first glance, this question seems silly. As we saw in the previous section, the motion of stars in a disk galaxy is governed much more by the gravitational field of the halo, not the disk. So why should perturbations to the gravitational field of the disk be important?

It is partly a matter of resonance. If a pattern of enhanced density actually rotated with a single angular frequency Ω_p, then this rotating pattern would produce a perturbation $\phi_1(R, \theta - \Omega_p t)$ in the gravitational potential. Since it must be periodic in θ with period 2π, it could be written as a Fourier series

$$\phi_1(R, \theta, t) = \sum_{m=-\infty}^{\infty} C_m(R) e^{im[\theta - \Omega_p t]}. \qquad (4.4.35)$$

A star on a circular orbit with radius R_0 and orbital frequency $\Omega(R_0)$ will experience a gravitational perturbation with a time-dependence given by a sum of terms proportional to $e^{im[\Omega(R_0) - \Omega_p]t}$. As we have seen, a star on a circular orbit of radius R_0 in an unperturbed axisymmetric gravitational potential has a natural mode of radial oscillation, with frequency $\kappa(R_0)$, and therefore responds resonantly to the perturbation (4.4.35) if

$$\kappa(R_0) = m[\Omega(R_0) - \Omega_p]. \qquad (4.4.36)$$

These are called *Lindblad resonances*. (There is also a corotation resonance, in which the $m = 0$ term in the gravitational potential causes a shift in phase of the star in its circular orbit.) These resonances have a finite width, because the actual time-dependence of any disk pattern is a superposition of terms with a range of pattern speeds, so orbits respond resonantly for radii for which $\Omega(R_0) - \kappa(R_0)/m$ is within this range of values of the pattern speed for some integer m. In particular, as we have seen, for a number of galaxies there is a rather well-defined pattern speed equal to the value taken by $\Omega(R_0) - \kappa(R_0)/2$ for a broad range of radii, so the resonance condition (4.4.36) will be satisfied for $m = 2$ and radii in this range.

The gravitational effect of perturbations to the disk density can also be enhanced by a phenomenon known as *swing amplification*.[30] In some cases, such as the winding up of a leading spiral into a trailing spiral, an instability similar to the Jeans instability discussed in Section 3.4 can greatly amplify the density contrast.

There is a large literature on this subject, going back to the cited work of Lindblad and Lin and Shu, and continuing with work of Goldreich and Lynden-Bell[31] and Julian and Toomre.[32] Much of this is discussed in the cited book by Binney and Tremaine. It is too complicated to go further into here.

4.5 Quasars

By the early 1960s a number of radio sources had been identified with optical objects that like stars have apparent sizes too small to be resolved, and hence were called *quasi-stellar sources*. In 1963, Maarten Schmidt[33] was able to measure the redshift of one of these sources, 3C273 (that is, number 273 of the 471 radio sources in the Third Cambridge Catalog). The fractional increase in wavelength of spectral lines from 3C273 was $z = 0.158$, indicating that this source is at a cosmological distance, about 600 Mpc. Judging from its apparent luminosity and its distance, the intrinsic luminosity of 3C273 was found to be enormous, over about $10^{12} L_\odot$, more than typical whole galaxies. Soon other quasi-stellar sources were discovered at even greater distances, such as 3C48 with redshift $z = 0.37$. Other objects with apparent sizes too small to be resolved that did not emit appreciable power at radio wavelengths were discovered optically at very large redshift, and have become known as *quasi-stellar objects*. In this section we use the term "quasar" to refer to both quasi-stellar sources and quasi-stellar objects.

Judging from the size of the extended regions of radio emission around some of the quasi-stellar sources, they seem to have been shining for at least 10^6 years, in which time they would have emitted an energy of order $10^5 M_\odot c^2$. And yet

---

[30] A. Toomre, "What Amplifies the Spirals?," in *The Structure and Evolution of Normal Galaxies*, ed. S. M. Fall and D. Lynden-Bell (Cambridge University Press, Cambridge, 1981), p. 111.
[31] P. Goldreich and D. Lynden-Bell, *Mon. Not. Roy. Astron. Soc.* **130**, 125 (1965).
[32] W. H. Julian and A. Toomre, *Astrophys. J.* **146**, 810 (1966).
[33] M. Schmidt, *Nature* **197**, 1040 (1963).

the luminosity of some of these sources fluctuates appreciably in a few days, indicating that they could not be much bigger than our solar system.

What could be producing this much energy in so small a space? Probably not nuclear processes. The conversion of hydrogen into iron produces about 8 MeV per nucleon, so nuclear processes in a mass M can't produce an energy greater than about $0.008 Mc^2$. To produce $10^5 M_\odot c^2$ by nuclear processes would thus require a mass greater than about $10^7 M_\odot$. A mass M that large in a region the size R of the solar system would have a gravitational binding energy GM^2/R of about $10^4 M_\odot c^2$, so gravitational condensation necessarily would provide much if not all of the energy radiated.[34] In order for gravitational condensation in such a small region to produce an energy of order $10^5 M_\odot c^2$, a mass even greater than $10^7 M_\odot$ is required. There is no known astrophysical system that contains so much mass in such a small region, other than a black hole.

Today it is thought that not only quasars but most or all galaxies including our own contain a black hole of mass $10^6 M_\odot$ to $10^{12} M_\odot$ at their centers. Quasars are just the galaxies whose central black hole happens to be surrounded with interstellar matter, which by accretion onto the black hole produces the enormous luminosity observed. The theory of this accretion is pretty much as described in more general terms in Section 3.5, with one significant difference. Instead of identifying the radius R_0 of the inner edge of the accretion disk as the radius of the central star, it is commonly assumed that for an accretion disk around a black hole, the inner radius R_0 is the minimum radius of any stable circular orbit around a spherically symmetric mass M, which is shown in the appendix to this section to be given by general relativity as three times the Schwarzschild radius, or $R_0 = 6MG/c^2$ in "standard coordinates," defined below by the metric (4.5.A11). With $R_0 = 6MG/c^2$, Eq. (3.5.29) gives a rate of viscous heating of the whole disk equal to $\dot{M}c^2/12$, where $\dot{M}$ is the rate of mass accretion. (Of course, it is only a crude approximation to use Eq. (3.5.29), which was derived in Section 3.5 from purely Newtonian physics, so close to a black hole that general relativistic effects are important.) Hence viscous effects in an accretion disk around a black hole can convert about $1/12$ of the mass flowing into the black hole entirely to heat. With this efficiency, the $10^{12} L_\odot$ luminosity estimated for quasars like 3C273 is produced by the infall of roughly one solar mass per year.

Appendix: Orbits of Minimum Radius Around Black Holes

In Newtonian mechanics, there are stable circular orbits of any radius about a point mass. As noted above in this section, this is not true in general relativity.

---

[34] This is essentially the argument of D. Lynden-Bell, *Nature* **223**, 690 (1969). This article presented the model of a quasar as a black hole surrounded by an accretion disk heated by viscosity.

4.5 Quasars

This can be shown by calculating the epicyclic frequency as a function of orbital radius, noting that for small radii the frequency becomes imaginary and hence the orbit becomes unstable. But of course, here we have to use the general relativistic equation of motion.

The general formalism based on the Principle of Equivalence yields the equation of motion for the radial coordinate r of a particle in a static spherically symmetric gravitational field:[35]

$$\frac{A(r)}{c^2 B^2(r)} \left(\frac{dr}{dt}\right)^2 + \frac{J^2}{r^2} - \frac{1}{B(r)} = -E. \qquad (4.5.A1)$$

Here $A(r)$ and $B(r)$ are two functions characterizing an arbitrary spherically symmetric gravitational field. (They are metric components in the "standard" spacetime coordinate system, $A(r) = g_{rr}(r)$ and $B(r) = -g_{tt}(r)$, but we will not need this information here.) Also, J^2 and E are positive parameters characterizing the various possible orbits, with no simple connection to the quantities denoted J and E in earlier sections. We will later take $A(r)$ and $B(r)$ to have the form required by the Einstein field equations, but for the present we will leave them as arbitrary functions.

For a circular orbit with a time-independent radius r_0, this gives

$$\frac{J^2}{r_0^2} - \frac{1}{B(r_0)} = -E, \qquad (4.5.A2)$$

so we can eliminate E, and write Eq. (4.5.A1) as

$$\frac{A(r)}{c^2 B^2(r)} \left(\frac{dr}{dt}\right)^2 + J^2 \left(\frac{1}{r^2} - \frac{1}{r_0^2}\right) - \left(\frac{1}{B(r)} - \frac{1}{B(r_0)}\right) = 0. \qquad (4.5.A3)$$

Next, consider a perturbation

$$r(t) = r_0 + \Delta(t), \qquad (4.5.A4)$$

with $\Delta(t)$ very small. To first order in Δ, Eq. (4.5.A3) gives

$$-\frac{2J^2}{r_0^3} + \frac{B'(r_0)}{B^2(r_0)} = 0, \qquad (4.5.A5)$$

so that the equation of motion (4.5.A3) may be written

$$\frac{A(r)}{c^2 B^2(r)} \left(\frac{dr}{dt}\right)^2 + \frac{r_0 B'(r_0)}{2 B^2(r_0)} \left(\frac{r_0^2}{r^2} - 1\right) - \left(\frac{1}{B(r)} - \frac{1}{B(r_0)}\right) = 0. \qquad (4.5.A6)$$

---

[35] For a textbook derivation, see S. Weinberg, *Gravitation and Cosmology* (Wiley, New York, 1972), Eq. (8.4.19). The radial coordinate is here denoted r because we are dealing with the case of spherical symmetry.

From Eqs. (4.5.A2) and (4.5.A5), we see that the conditions that J^2 and E are positive limit r_0 to values for which, respectively,

$$B'(r_0) > 0 \qquad (4.5.A7)$$

and

$$r_0 B'(r_0) < 2B(r_0). \qquad (4.5.A8)$$

We will see below that the requirement of stability sets a more stringent condition on r_0.

To find the limit on r_0 set by the requirement that the orbit be stable. we evaluate the terms in Eq. (4.5.A6) of second order in Δ:

$$\frac{A(r_0)}{c^2 B^2(r_0)}\left(\frac{d\Delta}{dt}\right)^2 + \Delta^2\left[\frac{3B'(r_0)}{2r_0 B^2(r_0)} + \frac{B''(r_0)}{2B^2(r_0)} - \frac{B'^2(r_0)}{B^3(r_0)}\right] = 0,$$

or in other words

$$\left(\frac{d\Delta}{dt}\right)^2 + \kappa^2(r_0)\Delta^2 = 0, \qquad (4.5.A9)$$

where $\kappa(r_0)$ is the epicyclic frequency

$$\kappa^2(r_0) = \frac{c^2}{A(r_0)}\left[\frac{3B'(r_0)}{2r_0} + \frac{B''(r_0)}{2} - \frac{B'^2(r_0)}{B(r_0)}\right]. \qquad (4.5.A10)$$

For $\kappa^2(r_0)$ negative the general solution of Eq. (4.5.A9) is a superposition of an exponentially growing term, proportional to $\exp(|\kappa(r_0)|t)$, and a decaying solution proportional to $\exp(-|\kappa(r_0)|t)$. The condition that there should be no exponentially growing solution for the displacement Δ is simply that the expression (4.5.A10) for $\kappa^2(r_0)$ should be positive, in which case the general solution for $\Delta(t)$ merely oscillates.

Now for the first time we will take $A(r)$ and $B(r)$ to have the form found from the Einstein equations in empty space outside a spherical mass M:

$$A(r) \equiv g_{rr}(r) = \left(1 - 2MG/c^2 r\right)^{-1}, \qquad B(r) \equiv -g_{tt}(r) = 1 - 2MG/c^2 r. \qquad (4.5.A11)$$

The squared epicylic frequency (4.5.A10) is then

$$\kappa^2(r_0) = \frac{MG}{r_0^3} - \frac{6M^2 G^2}{4c^4 r_0^4}. \qquad (4.5.A12)$$

This is positive, and the orbits are therefore stable at least against infinitesimal perturbations, if and only if

$$r_0 > 6MG/c^2. \qquad (4.5.A13)$$

With $A(r)$ and $B(r)$ given by Eq. (4.5.A11), the condition (4.5.A13) is more restrictive than the condition (4.5.A7) for J^2 to be positive, which is here automatically satisfied, and the condition (4.5.A8) for E to be positive, which only requires that $r_0 > 3MG/c^2$. Hence Eq. (4.5.A13) is the lower bound on the radii of stable circular orbits around a non-rotating black hole, whose event horizon is $2MG/c^2$. Black hole rotation would materially affect this limit.

Bibliography for Chapter 4

- G. Bertin and C. C. Lin, *Spiral Structure in Galaxies* (MIT Press, Cambridge, MA, 1996).
- J. Binney and S. Tremaine, *Galactic Dynamics*, 2nd edn. (Princeton University Press, Princeton, NJ, 2008).
- S. M. Fall and D. Lynden-Bell, eds., *The Structure and Evolution of Normal Galaxies* (Cambridge University Press, Cambridge, 1981).
- A. M. Fridman and V. L. Polyachenko, *Physics of Gravitating Systems*, trans. A. B. Aries and I. N. Poliakoff (Springer, New York, 1984)
  - Volume I: *Equilibrium and Stability*,
  - Volume II: *Nonlinear Collapse Processes*.
- S. Kato, J. Fukue, and S. Mineshige, *Black Hole Accretion Disks* (Kyoto University Press, Kyoto, 2008).
- D. Mihalas and J. Binney, *Galactic Astronomy* (W. H. Freeman, New York, 1981).
- H. Mo, F. van den Bosch, and S. White, *Galaxy Formation and Evolution* (Cambridge University Press, Cambridge, 2010).

Assorted Problems

1. Suppose that a spherically symmetric star is surrounded by a medium that exerts a uniform pressure p_0 on the surface of the star. Derive a relation between the gravitational energy Ω, the integral of the pressure $p(r)$ over the volume of the star, the star's radius R, and p_0.

2. Consider a variant of the CNO cycle, in which

   i : $^1H + {}^{12}C \to {}^{13}N + \gamma$
   ii : $^{13}N \to {}^{13}C + e^+ + \nu_e$
   iii : $^1H + {}^{13}C \to {}^{14}N + \gamma$
   iv : $^1H + {}^{14}N \to {}^{15}O + \gamma$
   v : $^{15}O \to {}^{15}N + e^+ + \nu_e$

   and either

   vi : $^1H + {}^{15}N \to {}^{12}C + {}^4He$ with probability P

   or else

   vii : $^1H + {}^{15}N \to {}^{16}O + \gamma$ with probability $1 - P$

   followed by

   viii : $^1H + {}^{16}O \to {}^{17}F + \gamma$
   ix : $^{17}F \to {}^{17}O + e^+ + \nu_e$
   x : $^1H + {}^{17}O \to {}^{14}N + {}^4He.$

   Express the equilibrium rate Γ per volume of each of these reactions in terms of $\Gamma(i)$ and P.

3. Suppose that the rate at which photons traveling in a direction $\hat{n}$ are scattered into a small solid angle $d^2\hat{n}'$ around a direction $\hat{n}'$ in stellar material of density ρ is $c\rho[\mathcal{A} + \mathcal{B}\cos\theta]\,d^2\hat{n}'$, where $\mathcal{A}$ and $\mathcal{B}$ are constants and θ

is the angle between $\hat{n}'$ and $\hat{n}$. How much does this scattering contribute to the total opacity?

4. Suppose that in a star the opacity κ and nuclear energy generation rate per mass ϵ are proportional to powers of mass density and temperature:

$$\kappa \propto \rho^\alpha (k_B T)^\beta, \quad \epsilon \propto \rho^\lambda (k_B T)^\nu.$$

Assume that the star is supported chiefly by gas pressure, with radiation pressure much smaller. Using dimensional analysis, show that the central density $\rho(0)$ of the star is proportional to a power D of the star's mass, and find D as a function of α, β, λ, and ν. What value would you expect for D if opacity arose chiefly from Thomson scattering and nuclear energy were generated by the CNO cycle, with suppression due to the Coulomb barrier by a factor of order e^{-48}?

5. In any star, we expect the density near the center to be given by an expansion in r^2:

$$\rho(r) = \rho(0)[1 + ar^2 + \cdots].$$

Give a formula for a in the case of a polytrope of index Γ.

6. Suppose that for the substance of some star the heat energy density $\mathcal{E}$ is related to the pressure p by

$$\mathcal{E} = Np^2,$$

where N is a constant. Find a function of p to which the mass density ρ is proportional during adiabatic processes. Use this relation to give a condition for the star to be stable against the onset of convection. Express this condition as a statement that stability requires some quantity to increase with increasing radial coordinate r.

7. Suppose that electrons had spin s instead of 1/2, so that the number of electrons that can occupy any orbital state is $2s + 1$ rather than 2. How would the mass and radius of a white dwarf star with a given central density depend on s? (You can limit the answers to white dwarfs that are well approximated as polytropes with $\Gamma = 5/3$ and $\Gamma = 4/3$.)

8. Consider a binary star with a circular orbit and period $2\pi/\Omega$, in which both stars have the same mass m. Give an equation relating the Cartesian coordinates x, y, and z for points on the surface of the Roche lobe (that is, the surface of constant effective potential, for a value of the potential at which the two parts of this surface surrounding each star meet at a point).

9. The ^{16}O nucleus is a spinless boson. Describe the rotational states of the diatomic O_2 molecule. Why is O_2 less important in cooling interstellar matter than carbon monoxide?

10. Assume that interstellar matter is in local thermal equilibrium with a constant ratio v_s^2 of pressure to density, but with an initial density $\rho(r)$ that has the (unrealistic) form

$$\rho(r) = C/r,$$

where r is the distance to a point at $r = 0$, and C is a constant. What is the initial radius of the smallest sphere centered at $r = 0$ that undergoes gravitational collapse?

11. Suppose that gravitation has a finite range $\mathcal{R}$, so that instead of the Poisson equation the gravitational potential satisfies

$$(\nabla^2 + \mathcal{R}^{-2})\phi = 4\pi G \rho.$$

- Find a solution of this equation and the continuity and Euler equations, for which the quantities ρ, p, and ϕ take values ρ_0, p_0, and ϕ_0 that are independent of position and time, and the velocity $\mathbf{u}$ vanishes.

- Consider a perturbed solution

$$\rho = \rho_0 + \rho_1, \quad p = p_0 + p_1, \quad \phi = \phi_0 + \phi_1, \quad \mathbf{u} = \mathbf{u}_1,$$

where quantities with subscript 1 are infinitesimal and have position and time-dependence given by a factor $\exp(i\mathbf{q}\cdot\mathbf{x} - i\omega t)$. You can assume that $p_1 = v_s^2 \rho_1$, with a constant v_s^2. Find the relation between ω and $\mathbf{q}$.

- Find the Jeans wave number, the maximum value of $|\mathbf{q}|$ for which this solution grows with time.

12. Suppose that in a collisionless gas, the third moment of the deviation of the velocity $\mathbf{v}$ from its mean value $\bar{\mathbf{v}}(\mathbf{x},t)$ vanishes. That is,

$$\int d^3v \, f(\mathbf{x},\mathbf{v},t)\bigl(v_i - \bar{v}_i(\mathbf{x},t)\bigr)\bigl(v_j - \bar{v}_j(\mathbf{x},t)\bigr)\bigl(v_k - \bar{v}_k(\mathbf{x},t)\bigr) = 0,$$

where $f(\mathbf{x},\mathbf{v},t)$ is the distribution function. Write a differential equation for the rate of change of the second moment $\Pi_{ij}(\mathbf{x},t)$.

Author Index

Aasi, J., 105
Abbott, P. B., 110, 111
Abdurashitov, J. N., 36
Abramowicz, M. A., 148
Adams, W. S., 125
Ahmad, Q. R., 36
Albalat, S., 168
Allen, C. W., 54
Andersen, J., 47, 48
Anderson, W. G., 116
Anselmann, P., 36
Archimedes, 56

Baade, W., 74
Bahcall, J. N., 36
Bahcall, N., 184
Barish, B., 105
Barlow, M. J., 125
Bell, S. J., 76
Bertin, G., 193, 203
Bessel, F., 1
Bethe, H. A., 31
Biedenharn, L. C., 166
Binney, J., 187, 194, 195, 198, 199, 203
Bland-Harwood, J., 185
Bode, P., 184
Böhm-Vitesse, E., 80
Bondi, H., 160, 161
Branch, D., 116
Bregman, J. N., 184

Brussaard, P. J., 168
Burnstein, D., 186

Chandrasekhar, S., xiv, 3, 61, 62, 80
Chiu, H.-Y., 24, 80
Chiu, I., 184
Clauser, J. F., 125, 126
Collins, R. A., 76
Coulter, D. A., 112
Cox, J. P., 16, 48, 80
Creighton, J. D. E., 116
Critchfield, C. H., 31

D'Onghia, E., 191
Davis, R., 36
Debye, P., 169
Dermott, S. F., 171
Detweiler, S., 112
Draine, B. T., 167, 171
Drever, R., 105
Dyson, J. E., 140, 171

Eddington, A. S., 176
Eggleton, P. A., 92
Einasto, J., 186
Einstein, A., 85, 119-121

Faber, S. M., 187
Fall, S. M., 203
Field, G., 125, 126
Flauger, R., 21

Flores, G. R., 187
Flores, R., 187
Ford, W. K., 186
Ford, W. K., Jr., 186
Frank, J., 171
Frenk, C. S., 184
Fridman, A. M., 203
Fukue, J., 148, 171, 203

Gamow, G., 32
Gaunt, J. A., 24
Gerhard, O., 185
Giuli, R. T., 16, 48, 80
Gnat, O., 134
Gold, T., 76
Goldreich, P., 199
Graham, J. A., 186
Grossan, B., 112

Hankel, H., 158
Hansen, C. J., 6, 48, 49, 80
Harmer, D. S., 36
Harrison, B. K., 81
Herbig, G. H., 125, 126
Hernquist, L., 191
Hertzsprung, E., 19
Hewish, A., 76
High z-Supernova Search Team, 186
Hildebrandt, W., 49, 81
Hitchcock, J. L., 125, 126
Hoffman, K. C., 36
Honma, M., 185
Hückel, E., 169
Hulse, R. A., 92, 93, 96
Hummer, D. G., 165
Hunter, J. H., Jr., 171

Jeans, J., 118, 145
Julian, W. H., 199

Kaasik, A., 186
Kalnajs, A. J., 196, 197
Kalogera, V. 110
Karzas, W. J., 165, 166
Kato, S., 148, 171, 203

Kawaler, S. D., 6, 48, 49, 80
Kelvin, Lord, 6 see Thomson, W.
Kenney, J. P. D., 186
King, A. R., 171
Kippenhahn, R., 54, 81
Kopal, Z., 92, 116
Kramers, H., 23, 24, 164
Kumar, P., 112

Landau, L. D., 76, 150, 155
Larmor, J., 5, 94
Latter, R., 165, 166
LeBlanc, F., 18, 39, 47, 81
Li, L.-X., 97
Lifshitz, E. M., 150, 155
Lightman, A., 81
LIGO Scientific Collaboration, 105, 110, 111
Lin, C. C., 191, 193, 199, 203
Lindblad, B., 196, 199
Lindblad, G., 196
Lynden-Bell, D., 190, 199, 200, 203

Maciel, W. J., 130, 167, 171
Maggiore M., 171
Magnus, W., 189
Matthews, J., 95
McKee, C. F., 145
McKellar, A., 125
Melia, F., 163, 171
Metzger, B. D., 97
Mihalas, D., 203
Mineshige, S., 148, 171, 203
Misner, C. W., 116
Mo, H., 134, 135, 143, 171, 203

Narayan, R., 110
Navarro, J. F., 184
Nice, D. J., 93, 96

Oberhettinger, F., 189
Omadaka, T., 185
Oort, J. H., 173
Oppenheimer, J. R., 76
Oster, L., 167

Osterbrock, D. E., 127, 167, 171
Ostriker, E. C., 145
Ostriker, J. P., 186, 187

Pacini, F., 76
Paczyński, B., 92, 97
Parker, E., 163
Peebles, P. J. E., 186, 187
Perlmutter, S., 186
Peters, P. C., 95
Pilkington, J. D. H., 76
Planck Collective XIII, 184
Polyachenko, V. L., 203
Pontecorvo, B., 36
Pooley, D., 112
Prialnik, D., 81
Primack, J. R., 187
Pringle, J. E., 159, 190

Raine, D. J., 171
Rasheed, B., 184
Research Laboratory for Electronics, 105
Riess, A. G., 186
Ritter, H., 49, 81
Roche, É. A., 88
Rosse, Earl of, 197
Rosseland, S., 14
Rubin, V. C., 186
Russell, D., 185
Russell, H. N., 17, 19
Rybicki, G. B., 81

Saar, E., 186
Salpeter, E. E., 40
Scheuer, F. A. G., 167
Schindler, S., 184
Schmidt, M., 199
Schwarzschild, M., 81
Scott, P. F., 76
Shakura, N. I., 157
Shapiro, I. I., 85
Shapiro, S., 66, 73, 74, 81
Shklovsky, I. S., 125
Shu, F., 81

Shu, F. H., 191, 199
Sofue, Y., 185
Sommerfeld, A. J., 166
Spergel, D. N., 110, 184
Spitzer, L., Jr., 131, 134, 137, 167, 171, 174
Sternberg, A., 134
Strömgren, B., 117, 126
Sunyaev, R. A., 157
Supernova Cosmology Project, 186
Sutherland, R. S., 165

Tanvir, N. R., 97
Tayler, R. J., 39, 81
Taylor, J. H., 3, 92, 93, 96, 110
Teukolsky, S., 66, 73, 74, 81
Thaddeus, P., 125, 126
Thomas, H.-C., 49, 81
Thomson, W., 5 see Kelvin, Lord
Thonnard, N., 186
Thorne, K., 105
Thorne, K. S., 81, 116
Toomre, A., 199
Tremaine, S., 187, 194, 195, 198, 199, 203
Trimble, V., 6, 48, 49, 80

van de Hulst, H. C., 168
van den Bosch, F., 134, 143, 171, 203
van Hoof, P. A. M., 165, 167, 168
Vikhlinin, A., 184
Virgo Collaboration, 110, 111
Vogelsberger, M., 191
Vogt, H., 17
Volkoff, G. M., 76
von Fraunhofer, J., 1
von Weizsäcker, C. F., 31

Wakano, M., 81
Watson, G. N., 158
Weber, J., 104
Weigert, A., 54, 81
Weinberg, S., 20, 21, 27, 28, 59, 68, 85, 98, 116, 119, 125, 133, 139, 150, 168, 201

Weisberg, J. M., 93, 96
Weiss, A., 49, 81
Weiss, R., 105
Wheeler, J. A., 81, 116
Wheeler, J. C., 112, 116, 171
White, S., 134, 143, 171, 203
White, S. D. M., 184
Williams, D. A., 140, 171

Wilson, R. E., 171
Wollaston, W., 1
Woolf, N. J., 125

Yahil, A., 186

Zimmerman, A., 168
Zwicky, F., 74

Subject Index

absorption of radiation, 7–8
  bound–bound transitions, 8, 26
  bound–free transitions, 7, 26
  free–free transitions, 7, 23–30
  net rate, 21
  spectral line at 21 cm, 124
  spectral lines from cyanogen, 125–126
  spectral lines from sodium, 125
accretion
  disks, 148–160, 200
  spheres, 160–164
acoustic velocity, 143–146, 157
advection, 156
atmospheres of stars, 16

baryon number, 66–67, 182–184
binary stars, *see* eclipsing binaries; equipotential surfaces; Hulse–Taylor binary pulsar; L1; Roche limit; Roche lobes; Sirius; spectroscopic binaries
black holes, 112, 124, 148, 154
  Cygnus X-1, 160
  M31*, 161, 164
  supermassive, 200
  see also coalescence
black-body radiation
  cosmic background, 125–126, 184
  Planck distribution, 12, 121–122, 130

Boltzmann equation, *see* collisionless Boltzmann equation
Bondi accretion, *see* accretion spheres
Born approximation, 28–30, 141, 166–168
bremsstrahlung, 140–142, 164–170
buoyancy, 56

central pressure, 3
close encounters, *see* violent relaxation
clusters of galaxies, 178, 182–185
clusters of stars, 14, 18–20
  see also Hertzsprung–Russell relation; main sequence
coalescence
  of black hole binaries, 98, 101, 110–111
  of neutron star binaries, 97–98, 100–101, 111–112
collisionless Boltzmann equation
  derivation, 172–173
  moments, 173–175
  solutions, 175–178
color temperature, 18–19
column density, 122–123
confluent hypergeometric function, 165
convection
  efficient, 57–58, 78
  energy flux, 54–57

convection (cont.)
  instability, 50–55
  mixing length, 56
  Schwarzschild discriminant, 53
cooling
  cooling function, defined, 132
  delayed radiation, 138–149
  prompt radiation, 132–138
  see also bremsstrahlung; molecular transitions
core radius, 182
cosmic microwave background radiation, see black-body radiation
Coulomb barrier, 32–33, 40–42

dark matter, 174, 178, 179, 182–185
Debye screening, 29, 169–170
density waves, 191
distorted wave Born approximation, 28
Doppler shift
  in binaries, 84, 93
  in interstellar matter, 120, 125
  neglected in stellar interiors, 9
  of stars in disk galaxies, 185–187

eclipsing binaries, 47–48, 83–84
Eddington formula, 176
Eddington limit, 20, 52, 148–149
effective surface, see surfaces of stars
effective temperature
  of accretion disks, 156
  of stellar surfaces, 18
Einstein A and B coefficients, 119, 121
ellipticity, 83
  decay, 95
emission of radiation, 9, 11
  emission measure, 129
  emissivity, defined, 164
  Hα, Hβ, 130
  Lyman α line, 126
  spectral line at 21 cm, 123
  spectral lines from clouds in equilibrium, 122–123
  spectral lines from non-equilibrium regions, 123–124
  see also bremsstrahlung; cooling function; Eddington limit; infrared radiation; nuclear energy generation; recombination
entropy, 57–58, 78
epicyclic frequency, 195, 202
equipotential surfaces, 88–89
η Carinae A, 77
Euler equation, 145, 173

Fermi statistics and momentum, 70, 74
flux of radiation energy, 10, 13

galaxies, 172, 173–174, 179
  bulge, 185
  disk, 185–186, 187–190
  halo, 186–187
  M31, 186
  spherical, 180
  spirals, 187, 191–199
galaxy clusters, see clusters of galaxies
gamma ray bursts, 97, 112
Gaunt factor, 164–170
giant molecular clouds, 144
gravitational binding energy, 3–4
gravitational waves
  blind spots, 111, 114
  chirps and chirp mass, 96, 106
  detection, 77, 110–112
  emission, 92, 100–104, 112–115
  energy and momentum, 102–104
  field equations, 99, 100
  harmonic coordinates, 99
  helicity, 100
  quadrupole approximation, 94, 102–104
  transverse-traceless gauge, 100, 112
  see also coalescence; GW150914; GW170817; LIGO and Virgo

Subject Index

GW150914, 110–111
GW170817, 111, 114

Hankel transform, 158, 189
Hartree approximation, 136–138
heating function, 131
Hertzsprung–Russell relation, 19, 47
HII regions, 126–131
Hulse–Taylor binary pulsar, 92–93, 95, 97, 105
hydrostatic equilibrium equation, 1, 178

infrared radiation, 5, 139
instability, *see* stability
internal energy density, 4, 10
inverse bremsstrahlung, *see* absorption of radiation, free–free transitions
ionization, 134–135
see also photoionization
isothermal distribution, 180–183

Jeans mass and radius, 143–146
Jeans theorem, 176

kilonovae, *see* gamma ray bursts
Kramers opacity, 23, 26

Landau–Oppenheimer–Volkoff limit, 76, 106
Lane–Emden equation, 61–62, 181
Larmor formula for electromagnetic radiation, 94
LIGO and Virgo, 105, 106–110
Lindblad resonances, 198
LISA, 112
L1, 90
luminosity, defined, 16

main sequence, 19, 42–50
 central temperature–mass relation, 46
 duration, 48
 luminosity–mass relation, 45, 48
 radius–mass relation, 45, 49
masers, 124
Maxwell–Boltzmann distribution, 25, 32, 129, 133, 142
mean free path of photons, 8–9, 12, 16, 128
minimum energy velocity distribution, 190
molecular transitions, 138–140, 144

Navier–Stokes equation, 150
neutron stars, 74–77
 mass limit, 76, 106
 rotation, 76
 see also coalescence; GW 170817; pulsars
NFW distribution, 184
nominal surface, *see* surfaces of stars
nuclear energy generation, 5, 11
 CNO cycle, 31, 36–38
 Coulomb barrier, 32–33, 40–42
 crossover between proton–proton chain and CNO cycle, 38–39
 nucleosynthesis, 30, 39–40, 97, 112
 power law, 30, 35, 38
 proton–proton chain, 31, 34–36
 see also Coulomb barrier; weak interactions

Oort limit, 173–174
opacity
 defined, 11
 power law, 22, 23, 26
 Rosseland mean, 14, 21, 25
 see also absorption of radiation; Eddington limit; Kramers opacity; scattering of radiation
optical depth, 16, 46–47, 119, 122, 124, 149
orthohydrogen and parahydrogen, 139

photoionization, 130
pitch angle, 194
Planck distribution, *see* black-body radiation

plasma frequency, 146, 170
Poisson equation, 145, 149, 169, 187–189
polarization tensor, defined, 100
polytropes, 4–5, 53, 57–59, 60, 65–66, 71–72, 75, 161–162, 178–180
  see also isothermal distribution; Lane–Emden equation
precession of periastron, 84–85, 93
PSR 1913+16, see Hulse–Taylor binary pulsar
pulsars, 76–77, 121
  see also Hulse–Taylor binary pulsar

quasars, 199–202

radiation energy constant, 14, 77
radiative transport equations, 9–10, 13–14, 43, 118–122
radio sources, 76, 92, 199
recombination, 123–124, 127–131, 134–135
red giant stars, 19, 39, 43, 89
ringdown, 110
Roche limit, 86–88
Roche lobes, 89, 90–92
Rosseland mean, see opacity
Russell–Saunders notation, 136

scattering of radiation, 9, 21–22
  see also Thomson scattering
Schwarzschild solution and radius, 65, 106, 200, 202
screened Coulomb potential, 27, 29
  see also Debye screening
semi-latus rectum, defined, 83
Sirius, 83, 88
spectral lines, see absorption of radiation; emission of radiation
spectroscopic binaries, 84
stability
  of general stars, 4, 63–64
  orbits around black holes, 200–202
  relativistic terms in energy, 64–70, 72–74, 79

  see also convection; neutron stars; star formation; supermassive stars; white dwarfs
star formation, 5, 18, 132, 143–148
Stefan–Boltzmann constant, 16
stimulated emission, 21, 25, 121, 124
Strömgren spheres, see HII regions
Sun, 3, 16–17
  central density, 16
  central pressure, 3
  central temperature, 16, 39, 46–47
  convective zones, 57
  Kelvin lifetime, 6
  main sequence lifetime, 48
  rotation, 76
  solar neutrinos, 36
  surface, 16–17
supermassive stars, 77–80
supernovae, 74, 77, 80, 90, 97, 145
surfaces of stars, 9, 15–16

Thomson scattering, 8, 22–23
time dilation in binaries, 85–86
transonic accretion, 162
true surface, see surfaces of stars
turbulence, 145

ultraviolet radiation, 126–127, 144

variational principle, 58–59
velocity dispersion tensor, 174
violent relaxation, 174, 182
virial theorem, 3–4, 141
viscosity, 150, 155, 200
Vogt–Russell theorem, 17, 20, 60

weak interactions, 31–32, 34, 39
white dwarfs, 70–74, 90
  Chandrasekhar bound, 72
  neutronization, 74
  Sirius B, 88
  stability, 74
wind equation, 162–163
winding problem, 195, 198
WKB approximation, 41, 193

X-ray emission, 89, 160, 183